THE HOMESTEAD BUILDER

THE

HOMESTEAD BUILDER

PRACTICAL HINTS FOR HANDY-MEN

SHOWING CLEARLY

HOW TO PLAN AND CONSTRUCT DWELLINGS IN THE BUSH, ON THE PRAIRIE, OR ELSEWHERE,
CHEAPLY AND WELL,

WITH

WOOD, EARTH, OR GRAVEL

COPIOUSLY ILLUSTRATED

BY
C. P. DWYER, ARCHITECT

TENTH EDITION

THE LYONS PRESS

All inquiries should be addressed to: The Lyons Press,
123 West 18 Street, New York, New York 10011.

Originally titled *The Immigrant Builder*
First published 1872 by Claxton, Remsen & Haffelfinger
Printed in Canada

10 9 8 7 6 5 4 3 2

Library of Congress Cataloging-in-Publication Data.
Dwyer, Charles P.
The homestead builder: practical hints for handy-men /
copiously illustrated by C. P. Dwyer.—10th ed.
p. cm.
Reprint. Originally published: New York: Hurst, 1872.
Includes index.
ISBN 1-55821-728-2
1. Country homes. 2. Wooden-frame houses. 3. Adobe houses.
4. Architecture, Domestic. I. Title.
TH4850.D8 1998
690'.8—dc21 98-4206
CIP

Dedication.

TO

ALL ADVENTUROUS MEN,

WHOSE AMBITION IS TO MAKE A HOME AND FOUND
A FORTUNE UNDER DIFFICULTIES,

*IN ANY LOCATION AND WITH WHATEVER MEANS
NATURE MAY SUPPLY,*

This Little Book is Dedicated,

BY THE AUTHOR.

FOREWORD TO THE 1998 EDITION.

HAVE you ever put up a small building by yourself? If so, you know how difficult it is, even with modern tools and materials. Now go back a century and imagine that you're a homesteader on the tall-grass Midwestern prairie, or on the semi-arid benchlands of Montana, or perhaps in the rain-soaked valleys of Oregon. You've followed the dream of the settler—you have the title to 160 acres of farmland, free and clear, so long as you live on it and break the ground to the plow. However, the first thing you absolutely must do is put up some kind of shelter, a house or a cabin or a hut, so you can survive that first long winter. There's no home center down at the corner, no power tools, and no electricity. You've got to put up that first house all by yourself, with hand tools and a wagon-load of store-bought materials hauled from the nearest railroad town, provided you've got some money. And if you don't have money? Then it's local materials, whatever you can harvest by yourself from your quarter section of land.

This was the daunting situation of the settler throughout the American and Canadian West during the great migrations of the second half of the nineteenth century. Political dominion over the vast Western territories could only be ensured by settlement, so the promise of free land was broadcast in the cities of the east and throughout Europe. Thousands answered the call, though few were prepared for the hardships and privations they would endure. A family with even a little cash might buy lumber and hire a country carpenter to help. A family without any money was truly on its own, afoot on the land that was most likely a half mile or more from the nearest neighbor. No problem, you say, we'll build ourselves a comfy log cabin. All very well if you've settled where trees grow, and you've got skill with ax and saw, plus strength to move and lift the heavy logs. However, over vast stretches of the West there are no trees, never were. What now, city man?

This was the void into which the Philadelphia architect C. P. Dwyer sent his first book, *The Economic Cottage Builder,* in 1850. Over the next twenty years Dwyer made a number of visits to the West to see what people were building and to work on building projects himself. He developed a healthy disdain for the "florid fancies" and "ornamental cottages" favored by the architectural press, and a deep appreciation for the problems facing the settler. In 1872 Dwyer brought his handbook up to date in the form of the little volume you now hold, published with the earnest desire to help the homesteader "shelter his family with comfort and economy combined." Originally titled *The Immigrant Builder*, Dwyer's volume of "practical hints to handy-men" was widely sold throughout Europe and North America, through at least ten printings before the turn of the century.

Dwyer's advice throughout this handbook is practical, down-to-earth, and sound. No wood? Build with sod, adobe, or gravel, because "earth, by compression, is capable of forming a ready and reliable material for building." No water? Locate and dig a well "the width of a flour barrel and sink a good sound barrel in it." Want windows? "You must be entirely governed by the size of your panes of glass."

Even today, should you become possessed by the urgent desire to build a cabin of your own, you can succeed by following Dwyer's prescriptions. Indeed, it is remarkable how well his instructions have withstood the test of time. He will lead you not only to an economical building, but also to a comfortable one where you can enjoy the ease afforded by your labors.

John Kelsey
March 1998

PREFACE.

THERE has always been a want of desirable information among Immigrants, to enable them to use their own labor to advantage, and to turn their small means to the best account. The consequence of this want has been seriously felt by thousands in the waste of their little stock, and the inadequacy of their efforts to procure for themselves and families even the rude conveniences which might pass for comforts; all owing to the want of the necessary knowledge which would lead them in the shortest and surest way to the attainment of their simple object.

It is a singular fact that there has been hitherto a total neglect of this want in the matter of the construction of habitations by these Immigrants themselves without the precarious assistance of the few rude mechanics, called "country carpenters," to be found occasionally in wide sections of newly settled country; and that, while the press is at present run upon by architects desirous of making themselves known by their florid fancies in *villa* and *cottage orné* designs, adapted only to men who can well afford to call in, and amply remunerate their services, without any reference to their book efforts, that the immense class of emigrants, as well as immigrants constantly seeking shelter in our wide-spreading country, should be wholly without any mental aid whatever, to enable them to economize their labor, their valuable time, and their tiny resources. So great is the increase of this tidal wave of emigration and immigration at present, and so likely is it to continuously swell, impelled by constantly arising influences, that the necessity for the information spoken of becomes hourly more imperative.

Years of residence in the Great West, and an actual experience in "the Bush," gives to the author of this little book those advantages of intimate knowledge so absolutely necessary to the conveyance of practical information in an intelligible form. To these he might add a knowledge of the various modes of cheap construction in those European countries where necessity is the superintending architect, and the immediate locality is alone to be looked to for the supply of the material with which to build.

In a vast country like this of North America, the differences of climate are not greater than are the varieties of requisites for construction, and the wants of materials for fitting shelter against the inclemency of the weather; and it is with a thorough knowledge of these that this treatise has been prepared. Since the publication, some twenty years ago, of the *Economic Cottage Builder*, the author has marked the numerous improvements and inventions which have presented themselves; and now takes advantage of these by culling all that is available to the present purpose of putting his readers in possession of the most ready and reliable information on the subject which is so interesting to them as being so conducive to their immediate convenience and comfort.

In the treatment of the subject, the author has classified the materials, as WOOD, EARTH, and GRAVEL; and these three heads embrace the various modes by which each is most effectively brought into practical utility. However, the reader will find it to his advantage to carefully examine each mode; for he will, most likely, find here and there an idea to be engrafted on that which more immediately concerns him.

In the construction of his future home, it surely is the interest of the immigrant to carefully consider his requirements, and to so plan his house that he may have as few omissions or mistakes to regret in the future as possible. Let him on no account be led away from the dictates of his own judgment by the ignorant prejudices of old settlers; or, for the mere sake of hastening the accomplishment of his task, to pattern after some already built-up pile of discomfort. But, rather, to weigh carefully the consequences to the well-being and ease of his family, and his own satisfaction.

An error hastily built up cannot easily be remedied, but must remain as a lesson to impatience and thoughtlessness.

The plans suggested in the following pages are so arranged as to give present accommodation, and to afford an easy opportunity of adding to it in the future without interfering with the construction already in use. Thus, a fresh erection with its gable end to the rear wall having a short porch or covered passage between, would give the plan of the whole the form of a T, and insure perfect light and ventilation.

The barn should never be connected with the house, but have a sufficient yard between, in the centre of which the pump or well might be located. The barn, being very liable to destruction by fire, is too dangerous to be a very near neighbor of the dwelling; and for other reasons, it is more desirable that it should stand apart.

In conclusion, it may not be amiss to add, that, although

no effort is made in the following designs to introduce any tasteful effects, yet such can be at any time produced by the aid of a little judgment in simple ornamentation and calling in the ready assistance of Nature, with her flowers and her creepers to twine around and halo this home of the handy-man.

C. P. DWYER.

Philadelphia, 1872.

CONTENTS.

INTRODUCTORY.

THE pioneer who enters on his task of "settlement" is pretty much in the same condition as were the primitive inhabitants of this now wonderfully enlightened world of ours, when driven to eke out a living and a lodging where neither was readily to be found.

Writers on architecture have given us what, at best, are but their own crude imaginings of the modes adopted. But as the modes actually followed must have necessarily depended on the peculiar material presented for construction in the very various localities into which men entered, we can scarcely acknowledge this solution of the problem.

Be this as it may, our object now is to make use of all the available hints which science and experience give, to meet the immediate wants of the poorly provided immigrant builder, to enable him to construct his little homestead comfortably and permanently, protected from the inclemency of the weather and the inducement to disease.

But, although the pioneer class who seek their homes in the dense "bush," or on the distant plains, is a very large and greatly increasing one, there is still another and scarcely less extensive community to whom the wide outskirts of our overcrowded towns and cities offer inducements to settle themselves, and be their own landlords. To all such the information here presented may, in some one shape, be found suitable to their object and their ends.

Every man of small means having a family is naturally desirous of providing a shelter for them, and in so doing he desires to have comfort and economy combined.

Economy is nothing more nor less than the keeping within one's means — so that the amount of comfort to be obtained must, of necessity, be limited to that extent.

Now, what is meant by comfort, as concerns such a man, is a simple combination of *convenience* and *durability*. Convenience must be acquired by the sacrifice of effect or appearance, if necessary, for both cannot be at all times combined, especially in the economized space we are about to speak of; and durability is a quality which it behooves the poor man to be particular about, for he cannot afford a repetition of cost, as the rich man can. It simply means strength, and that strength is to be obtained, not by heavy material, (for economy must be kept in view,) but by judicious arrangement or combination. Thus, a light frame is very often twice as strong as one much heavier, for the reason that the combination of parts in the one is superior to that in the other. Here, then, we have economy of material producing strength at the same time that we secure lightness, and, with all, durability.

In the following hints, therefore, we will be guided invariably by these considerations, and will propose nothing to the reader but what we conscientiously believe to be for his interest to adopt.

As this little treatise may possibly fall into the hands of a variety of persons, each desirous of information, yet having a great diversity of material to work with, it will be advisable to endeavor to reach the wants of all, so far as possible. We will, therefore, treat of the different localities in which cheap dwellings may be called for.

Where the vast prairie stretches out towards the setting sun, timber is as scarce as it is plentiful in the thickly-shaded *bush*. There then is a necessity for other than wooden buildings. On such a site the poor man will find his most economic mode of building to be that which can be most easily executed with the material at hand; and that material he will readily find by digging down into the site of the house he desires to build. There are three modes of construction by which he can use the material thus presented to him. The loam, or richer coat, he can set aside as top-dressing for his vegetable garden, and make his walls of the clay or sub-soil, in the way we will direct in speaking of *adobé*, or *sun-dried brick;* and of *pisé*, or *compressed earth.* Should limestone be convenient, and a gravel-bed within reach, he will be able to adopt, if he chooses, the still better mode of construction called *concrete*, or *gravel-wall building.*

Where timber is plentiful, walls or enclosures in that material will prove desirable, as being more ready to the hand of the man impatient for a shelter as well as for him who desires to build cheaply; and as timbered land is most generally selected (where there is a choice), and in such localities saw-mills are surely to be found, the primitive *log hut* may afford a comfortable, if not a convenient home; or the more scientific *frame* give a house of very superior neatness, but at much greater outlay.

The *log-house* usually found in our backwoods is only remarkable for its roughness of construction, although often a picturesque feature in a landscape. But its interior, which ought by all means to be its best, is too generally its worst side; for there, where walls ought to present something approaching an even surface, the contrary is the rule. Everything is harsh, uneven, rough, uncouth. That they could be furred, lathed, and plastered, and thus rendered much more comfortable, is true; but that such a treatment is very seldom applied to them, shows that they are not esteemed worthy of the expense. This is a very improper conclusion; for any shelter which is deemed desirable for present purpose is certainly entitled to every attention that can render its means of comfort more complete, even as a temporary homestead for a struggling family. It is very true that the old log-house is pre-doomed to be the stable or cow-house; but that is no reason why it should be permitted to remain uncomfortable for the years it may be used as a dwelling for human beings, who possess a laudable ambition beyond its humble limits.

PRELIMINARY ADVICE.

THE Immigrant who goes South or West from our crowded cities to settle, and the Emigrant from Europe who comes to our hospitable shores to seek a home and independence, will alike find desirable information in the following pages; which treat exclusively of economic modes of construction, whereby the mere novice in building can erect his own homestead in the best manner that his limit of means will allow, and use to advantage the material which the location presents to his hand.

Unlike a book of ornamental cottages intended to adorn the suburbs of some city, the designs here presented are simple in the extreme, and look to comfort and permanence rather than to effect of appearance. Yet the humblest hut can with a little taste, practised in leisure hours, be made to put on beauty well adapted to its calling.

The utmost reliance can be placed on the advice given, for the author has himself executed constructions in each and every mode of building herein spoken of. The plans are all drawn to one scale.

Let the emigrant or the immigrant provide himself with the tools recommended in this book, and get them in the place where they are cheapest (as in the Eastern cities), and when leaving the last village, on his final stage to his destination, let him procure his window-glass, paint, and putty.

Let him read and study his wants, and go forward prepared.

LIST OF TOOLS TO BE BOUGHT.

Two saws.
Two axes.
Three hammers.
Three augers.
One or two gimlets.
Three or four files, different sizes.
Two planes (a jack- and a smoothing-).
Three chisels (inch, half, and quarter).
Two trowels.
One iron square.
One two-foot rule.
Two spades.
One or two iron rakes.
One oil-stone.
One glue-pot.

To these tools may be added a keg of eight-penny nails, and a keg of mixed four-penny and six-penny; a dozen sheets of sand-paper.

Many of the above may not be wanted immediately, but they will each and all prove useful at some time.

The steel and iron tools should be coated with oil, when put away, to preserve them from rust. Where two or more families go in company to settle in the same neighborhood, it would be well to make the collection of tools a joint-stock concern, until such time as each would be able to purchase independently.

Where there is a family, it is well to leave the helpless ones to board at the nearest place to the intended homestead. Any farmer will receive them on reasonable terms, and wait for remuneration, or take it out in labor. The men should go at once and carefully prepare the home for them.

LOCATING THE SITE.

Water.—Should there not be a spring or stream of drinking-water near which to locate, it will be necessary to sink a well. Great care must be taken to ensure a good, reliable quality of water. When a desirable spring is reached, the well should be formed. The readiest way to do this is to dig a hole the width of a flour-barrel, and sink a good sound barrel in it. Proceed to dig down below it, and force down the barrel by placing another upon it and digging under. So continue to dig and force down barrel after barrel until the required depth is gained. This is not a permanent well-lining, but it will serve a good purpose until such time as improved circumstances give a more reliable one.

Very great caution should be used in the making of wells, as poisonous gases sometimes collect in them, even in a night, making it fatal to life to venture into them.

It would be a good precaution to lower a lighted candle into the new-made well each morning, before a man ventures down into it. If the light is extinguished in its descent, then the well is dangerous, and before work is proceeded with, it would be a good practice to throw into it a bushel or two of quicklime. This will absorb all the dangerous gases in the well without fail, and render it safe to work in after an hour or so. However, it would be advisable to use the lighted-candle test once more, to ensure safety to those who risk the descent.

Water is often met with at ten or twelve feet deep; but it is most generally what is called a land-spring, and may not hold out in sultry weather. It is better, therefore, to go still deeper, until a permanent supply is secured. The flavor of water should likewise be tested from time to time until a perfectly palatable quality is found.

To reach the main spring of a really good water is a most desirable object; especially when we reflect that the want of it will prove so great a deprivation as to oblige the tenant to either quit his newly-built house, or suffer all the inconvenience of drawing water from a distance.

The first water reached, if palatable, (although it be a land-spring,) should be used until another and more permanent well is sunk.

Raise the ground around the top of the well, to prevent dirty water running into it, and fix a firm fence around, to prevent cattle or children falling into it. Do not cover it over, as water is all the purer for having the open air above it.

The pump, although the readiest, is not the best mode of raising water, as the suction disturbs the sediment and matter at the bottom, and, moreover, the well has to be shut in from the purification of the atmosphere. The bucket is more desirable; and to facilitate its working, it would be advisable to have two buckets attached to a continuous rope. Thus, when one is coming up full, the other is going down to be filled, and a saving of time and labor is the consequence. The axle on which the rope is to wind may be easily made out of the straight trunk of a tree, freed from its bark, cut to the required length, and having hard-wood pins inserted in its ends, to work in two holes in firm, upright pieces connected with the fence around the mouth of the well. If iron pins can be had, so much the better; and if iron rings for the ends of this axle could be added, the whole would prove more desirable. A handle must be fitted on to the right-hand pin (which for that purpose should be square); ash or oak would be the most proper material for this handle.

If, however, a pump is deemed necessary, allow no more than twenty-four feet from the surface of the water to the spout or nose of the pump, as any greater length of shaft will weaken, if not destroy, the power of suction. In fact, it would be surer to allow but twenty feet, particularly in rudely made pumps, such as would be most likely to be had in scantily populated places.

Drainage. — There is nothing easier to provide for at first, than the perfect carrying off of waste water and slops from a cottage; and yet, is there no more common occurrence than the neglect of it? It is true, that everything is hurried in the desire to secure an immediate shelter; but this is a matter so intimately connected with the health and cleanliness of the future dwelling, that there is no excuse for its being overlooked. Besides, the saving of waste water, or sewage, for manuring purposes, is very desirable. Let there be formed a large sewer sufficiently near the house to drain into it; and let such sewer be below the level of the cellar-floor. Make a cess-pool at the near end of it, and let the far end empty into the dung-pit.

As the site of the house should occupy the most elevated ground on the lot or place of location, it will be easy to give this sewer a good fall or inclination, which in any case it should have. A sewer should on no account be less than ten inches in width. Its simplest form is that of a deep V, and the ground should be dug out in that shape, and the sides be well tamped, or rammed. Boards should then be laid on these sides, and be made to come close together at the meeting-point of the V, whilst they are kept apart by sticks at the wider or upper end. The sewer should then be covered over with short slabs laid crossways; and these again by other slabs laid lengthways. Then the whole length may be covered in with rubbish, loose stones, and clay, up to the surface of the ground. When well done, this sewer will answer a good purpose, until improving means shall enable the owner to procure earthen pipes, drain-tiles, or to build a brick barrel-drain.

The cesspool should be sunk not less than six inches below the bottom of the sewer or drain, near the well or pump. It may be, say twelve inches wide; and it should have a flag stone or slate, a tile, a wooden or an iron partition, set upright in its centre, flush with the cover at top; but leaving at the bottom a space sufficient for the escape of the water. This divides the cesspool into two chambers, and effectually cuts off all disagreeable smells, as well as the inroads of vermin into the house from the sewer. At the farther end of the sewer it would be well to dig out and board a tank, or sink a couple of barrels, for the reception of the liquid manure. Over this should stand the privy; and this necessary little outbuilding should be so located or placed, that there may be ample opportunity for emptying the vault or tank from behind it without disturbing the privy, or any part of it.

CONSTRUCTION IN WOOD.

THE PRIMITIVE LOG-CABIN.

LAYING out the Lines.—Having previously cleared and levelled the site, the next thing in order is the marking out the exact size and position of the intended house; and this is done by first determining the front line, bearing in mind that the entrance should face southeasterly, if possible; thus securing the greater portion of the sun's influence on three sides, and a fair share on the fourth, or northwest. If not possessed of a compass, it will be very easy to watch the course of the sun, and be governed by it. In laying out the lines of the house, the first object should be to have the walls *perfectly square;* and this is effected by means of three rods respectively 6, 8, and 10 feet long. These are tacked together securely, in the form of a right-angled triangle, in the manner here shown. (Fig. 1.)

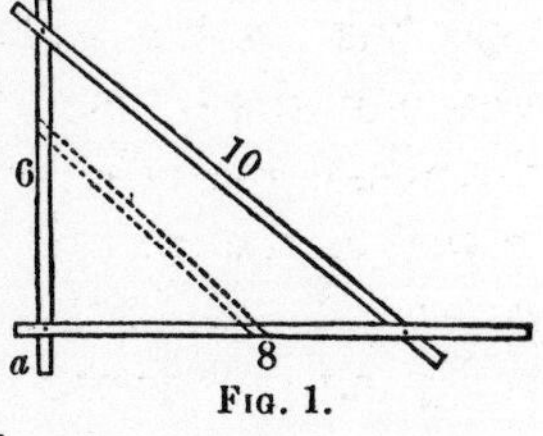

FIG. 1.

The dotted lines show a brace which would considerably strengthen the construction.

This square should be laid on the ground at any one of the corners of the proposed house. Either sides (6 or 8) should lie against the front line already marked out; and a string fastened to the nail at *a*, and longer than the greatest length of the house, from out to out, should be stretched both ways; and thus might be easily marked with stakes driven firmly into the ground, and having a nail driven in their tops around which to wind the string, the front and side walls made perfectly square with each other. The same course pursued at the other end of the front line, will be sufficient to accurately determine the position of the walls; and from these lines the exact lengths of the tree-trunks to be used may be decided on, and the cuts made by which they are to be halved together at the angles. These cuts are usually made with the axe; but a good saw will give a much neater and more accurate joint to the halving, using the chisel or axe to clear out the cuts. There will be no necessity for pinning these joinings together, as they can be made to fit tight by wedging, and their weight will keep them in place. The boring of holes and cutting of pins will thus be saved. Before making the cuts, chalk off the width or span of the log which is to cover that which you cut; and cut each log exactly alike, so that there may be no looseness in the fit.

Walls.—There must necessarily be an interspace between the logs which form the walls, as seen in the illustration, and the problem is to fill this up to the greatest advantage. Perhaps the best method of accomplishing the object in an economical manner, and at the same time to thoroughly secure it from the action of the weather, is to fill the space (whatever it will be) with a slab or split-trunk, laid lengthways, between every two trunks of timber.

This may afterwards be slushed or plastered up with mud prepared for that purpose, and will make a permanent job.

The system most generally in use, is to insert or wedge in chips, and plaster up. But it must be evident that this does not make so workmanlike a job as the one here proposed; for the mud-plaster, in drying, will shrink, and consequently crack and fall away. Besides, the slab or split-trunk, by being the same width as the wall, will form a furring on the inside to nail any required woodwork to, such as shelves, partitions, &c.

The logs or tree-trunks, where cut down, should be trimmed and notched as required, and then carried to the site of the intended house, where they are to be placed conveniently to the side or end on which they are to be laid in constructing the wall. Those intended to form the first row, nearest; the second row, next to the site of the wall; and so on out to the capping or last log.

It is always well to drive down short sharp-pointed piles at the four corners (say four at each corner), and to fill in between them with large stones and gravel. On these, cross-sleepers should be laid to receive the wall-logs; settlements will thus be avoided, and the house be secured from that great cause of ruin to this mode of building.

Previous to laying up the walls, the whole should be made perfectly level for the first course of logs to rest upon.

The Chimney. — The location of the chimney being determined, it will be necessary (if it is to be of stone) to drive down piles, and fill with gravel and broken stone in a secure manner; and to cut off those piles level with the surface of the ground, carefully covering them over with a thick course of stones, gravel, and sand, well pounded together, without any moisture. As rats are apt to seek the heat of the fireplace in winter, the hearth is their natural burrowing-place; and this precaution will, therefore, prove an obstacle to their inroads, as well as a firm and reliable foundation for the chimney.

We are now supposing the existence of stones on or near the premises. Of course, if such be not the case, the chimney will have to be constructed in a manner the least liable to accident by fire which can be thought of. This is not by any means an easy matter; and it, therefore, calls for mature consideration on the part of the builder. Probably the readiest mode is to set up four posts at the four corners of the intended chimney, and brace them well with slabs all around; then fill the box, so formed, with earth, and gravel, and sand (if convenient), ramming down the whole at every three or four inches in thickness, until a perfectly solid mass is made, say five feet high, six feet wide, and three feet deep. When the whole block or mass is complete, then remove the front slabs; that is, uncover the side which is to be the fireplace, and cut out carefully, with a spade, shovel, or other tool, the chimney opening, or fireplace, all the way up, leaving not less than twelve inches thickness of jamb on each side, and twelve inches thick of backing; so that the fireplace will be four feet wide and two feet deep. (Fig. 2.) Across the front of this fireplace, at say four feet high, will be laid a good sound log of oak or hickory, resting on the jambs on either side. (Fig. 3.) This lintel, or brest-summer, should be hewed square, and be eight inches thick on its flat side, by ten inches high; and be six feet long, or the full width of the chimney-breast, from out to out. Less dimensions may be used, but not with safety, as the brest-summer has to carry the weight

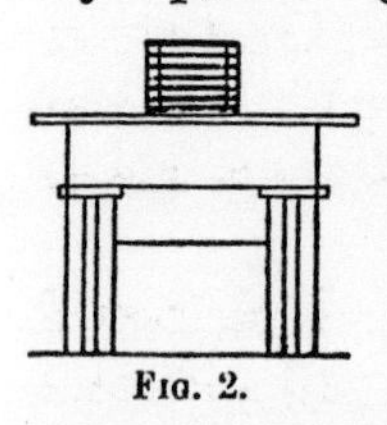

Fig. 2.

of the chimney. However, it need not be one piece, but two or three, to make the thickness. It should be carefully selected from the trees that are least combustible, such as bass-wood, oak, box-wood, or *lignum vitæ.* There is in every locality some wood that will suit this purpose. It would be well also to have the posts which compose the framing of the chimney-breast, of the same wood.

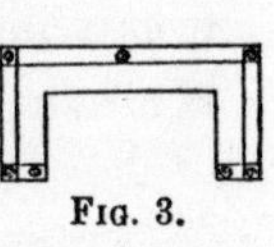

Fig. 3.

On the lintel may be nailed the mantel-shelf, sufficiently wide for the front of the chimney-flue, to rest on the inside of it. Wall-pieces, say four inches wide and two inches thick, should be laid in the centre of the chimney-jambs for the lintel to rest upon, (as shown Fig. 4.)

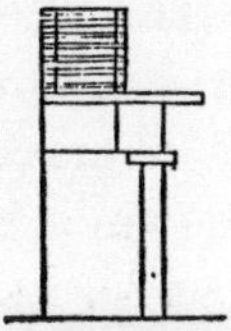

Fig. 4.

The flue should be narrowed in by means of slabs laid at intervals; and the earth built on them, and each course projecting inward an inch or so. The *throating* of the chimney is thus formed, and when it is eighteen by twenty-four inches, the flue may be continued up straight, until it reaches two feet above the ridge of the roof. The flue may be constructed of straight pieces of branches, say three or four inches thick, notched down on each other at the angles, well plastered inside and out.

This plaster should be made of clay in which gravel is mixed, and the whole be puddled with just sufficient water to make it into a stiff mortar. It should be well beaten with flat sticks, and put on with a board ten inches square, having a wooden or leather handle nailed on one side of it. The inside of the flue should be *pargetted* or plastered with fresh cow-manure. This grows very hard, and is not liable to crack when dry.

It also makes a smooth surface to the interior of the flue, and prevents the accumulation of soot and the consequent interruption of the ascent of the smoke.

It would be very advisable to cap the flue with a box, of the proper size, having hinged doors on the four sides; each door being connected to the one opposite to it by a stick or stout wire longer than the flue is wide, for the purpose of giving free passage to the smoke under all states of the wind.

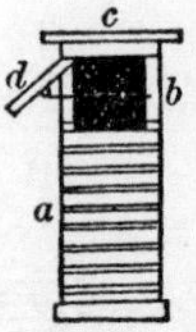

FIG. 5.

In Fig. 5, *a* represents the chimney-top; *b*, the box, or ventilator; *c*, the cover; *d d d*, the doors (of which there are four), all hinged to the under side of the cover. If iron hinges are not handy, leather or stout canvas, doubled, can be made to answer the purpose.

This ventilator should be securely put together, and carefully fastened in its place. It will be found very useful in guarding against "down-draft," or the driving back of the smoke by the wind; for, no matter from what quarter the wind comes, it will shut the door it strikes against, and push open the opposite door, as will be seen by the illustration. This ventilator will also serve to prevent the rain falling down the chimney and putting out the fire.

Of course, it is only presented as a rough substitute for those more perfect ventilators guarded by patent rights, and out of the reach of the pioneer. But it will be useful in its way, and serve until improving fortune enables the immigrant builder to better his condition.

The Roof. — The best mode of roofing is that which tends most to bind the building in its length as well as its breadth. For this purpose, it would be well to have a ridge-pole of good, hard wood (oak, if possible), not less than six inches diameter, stretched from gable to gable, and projecting a couple of feet beyond each gable. This pole should be sustained throughout its length by the forked rafters crossing each other under it and tied together with tough switches. In order to avoid the dividing of this ridge-pole, it would be well always to keep the chimney to one side of it. Across the house, from front to rear wall, heavy beams should be laid projecting, say four or five feet beyond the face of both of the walls; and these should be notched down on the walls, and also pinned with oak or ash pins, cut square, and driven into round holes somewhat smaller than they are, so that the hold of the pin will be tighter. These cross-beams will give great strength to the construction, besides forming the joists for an attic floor. Two feet apart, from centre to centre, will be amply sufficient. The rafters forming the roof should be made to stand upon them in this manner: Along the ends of the joists, just spoken of, lay eaves-poles as straight as can be had. Notch and pin these down very firmly to the joists; and then notch and pin the ends of the rafters to these eaves-poles. Divide up, or space the rafters, so that there will be a pair at every joist. Building in the bush, there will be no want of material wherewith to do this work thoroughly and well. Slabs will be perhaps the most convenient covering; but if the roof can be boarded, shingles would be the best roof for a log-house. If, however, there be a marsh or a pond near, on which reeds, sedge, or long, wild grass is plenty, it would be well to thatch the roof with such material, in preference to going to work to make shingles. Straw is still better for the purpose; but the other (if to be had) is the more economical.

Brushwood is used where the other cannot be conveniently had.

If the covering is to be of slabs, it is best to lay a first covering lengthways (that is from gable to gable), with the flat sides turned up. Make the edges even and fit them close. Next, nail on these slabs securely to the rafters; and then cover them crossways (or in the line of the pitch of the roof) with slabs, also made even on the edges; and nail all tight, having this time the flat sides turned down. Slips, two or three inches broad, nailed over the joints, would make the whole very complete, as well as give a good appearance.

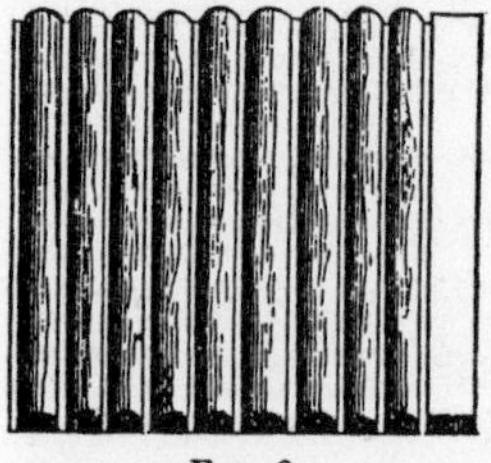

Fig. 6.

It must be observed that (Fig. 6) the gables of the roof are to be covered by plank, bearing on short cross-pieces laid on the wall for that purpose, and on line with the projecting eaves-poles, front and rear, and the ridge-pole at top of roof. Outside of this plank, and nailed on to the ends of the eaves-poles and ridge-pole, a nice even bough of spruce or pine would give a neat finish to the gable, besides protecting this gable-covering from the driving of rain or snow.

Shingles are laid on in the same manner as slates. A double course is laid along the eaves; and four inches up from the lower edge of that course is laid the second; and four inches up from that, the third, and so on to the top. In order to ensure these courses being perfectly straight, a well-chalked string is secured to a nail, driven temporarily at one end of the course, and when stretched to the other end and secured, it is lifted by the finger and thumb, and suddenly let go again, when a chalked line is left along the course to be followed; and this operation is to be repeated every time a fresh course is to be commenced.

Thatch. — This covering for a roof has at least the advantages of being able to keep out the heat in summer and the cold in winter, as well as being economical; but it is not, after all, a very desirable roofing material. Vermin find too ready a refuge in it, and it is very liable to take fire from the sparks flying from the chimney. The damp striking deeply into it, too, renders it a very unhealthy covering for a human habitation. However, as it is sometimes likely to be the *handiest* covering for a roof, we will here give the mode of constructing it:

In thatching, splints, made of small poles split in halves, are nailed upon the rafters. The straw, reeds, brushwood, or whatever else is to be used as a covering, is well wetted, and then put upon them; and, after being laid straight and combed with an instrument like the head of a rake, the upper end of the straw is tightly bound to the splints and rafters with rope-yarn, cord, or any convenient binding.

The manner in which thatch is laid upon a roof is different from other roof-covering, although, like them, it is begun at the eaves, or lower part of the roof; but, instead of being continued from one end of the roof in courses, the thatcher lays on as much as he can well reach across, and proceeding with that width from the eaves to the ridge, works on in that way from one end of the roof to the other. The ridge and sides of the roof are secured against the high winds, which are apt to damage them, by running small stakes in the way of skewers into and through them, then binding them tightly together with cords. The sides and the eaves are then clipped even with a pair of shears. Thatching, when well done, gives a very neat and comfortable appearance to a house; but it requires frequent repair to ensure its being perfectly weather-proof.

Eave-Troughs.—The very poorest log-cabin should have the means of conveying the rain-water from the roof; and this is a matter so easily accomplished that the discomfort and inconvenience attendant on the want of it need never be allowed to exist.

The most simple form of a trough or gutter for this purpose is made by cutting a smooth board of about nine inches wide down the middle with a saw, only leaving one part as much wider than the other as the thickness of the board. The edges should then be planed, and the edge of the narrow one set against the side of the other, near its edge, and there nailed; the side of the one into the edge, or thickness, of the other. If paint, or any substitute for it, can be had, it would be well to coat this joint previous to nailing up. Each end of this V-gutter is stopped up by nailing a piece cut to fit in between both sides. One end is, of course, kept highest when fixed under the eaves, so that the rain may have a fair current or fall, to the end at which it is to discharge into the spout.

The *spout* is made of four boards, four inches wide (or even three inches would answer), nailed together at their edges, so as to form a square tube through which the rain falls in its passage from the eaves-gutter to the barrel, or cistern, prepared for its reception. Every drop of rain-water should be saved, as it sometimes happens that wells will dry up in summer, and then a supply of rain-water would come handy to the household.

The top of the spout should be cut, so as to let the

eaves-gutter fit down into it; and a hole should be cut in the bottom of the eaves-gutter directly over the spout. The eaves-gutter should be nailed down to the top of the spout, and be supported every three feet on wooden brackets made out of boards, and cut V-shape to receive it. These brackets should be secured to cleats nailed to the walls, and should be long enough to keep the gutter at a proper distance, yet convenient for the drainage of the roof. All the stuff for the eaves-gutter, spout, and brackets, may be got out of inch-boards.

Flooring. — The house roofed and covered in, it will then be the proper time to attend to the carpentry and general finish of the whole. First, the flooring is to be laid; then the ceiling and partitions; and while these operations are going on, the windows and doors can be set in their places, having been previously made or bought.

It is usual to lay the flooring joists from front to rear wall, and where the chimney is in or near the middle (as it should always be), to leave out a joist or row, and set cross-joists, or, as they are called, bridging-joists, from one to the other; thus insulating the chimney, and preventing accidents from fire. These joists should be well-chosen, straight trunks of spruce, pine, or any convenient wood, eight to ten inches in diameter, according to the breadth of the house; and they should be laid not less than two feet, from centre to centre, apart. To lay them one foot apart would be still more desirable, as a stiff floor adds greatly to the strength of the house, and the flooring boards need then be no more than inch thick, whereas in the other case they should of necessity be inch and a half, or even more.

A rough floor, but a substantial one, may be made of

slabs, provided the spaces between the joists are filled up with earth or gravel, so as to have the round part of the slabs supported on a bed throughout, while the flat face turned up is made as level as possible. This floor, when covered with a rag-carpet, having a thin layer of straw under it, answers as well as any other. Of course, the slabs are to be stoutly nailed, or oak-pinned, at their thickest part to the joists.

The ceiling must be open, as boarding or plastering cannot be had recourse to — economy being the main object. An attic-loft, or second floor, can be easily constructed of boards nailed across the tie-beams, or poles, which cross the house from front to rear wall; and it would add very much to the comfort, if splints (two inches, or so, wide) were to be first tacked on, before laying the floor, that the joints between the boards may thus be covered, and dust and dirt be prevented falling from the loft. These splints would likewise add greatly to the stiffness of the floor. They can be readily formed by splitting up boards to the required width.

Partitions are constructed most readily of thin boards nailed to slats, secured to the floor below, and to other slats made fast to the ceiling-joists above.

As the chimney will occupy the middle of the house, it will divide the partition and lessen its extent. One part, or section, of this partition should be built on a line with the front of the chimney, thus leaving a recess for the stairs, or ladder to loft, to occupy. The boards forming the partition should be made even on the edges, so as to make as close a joint as possible. When they shrink, strips of strong paper should be pasted over them; or, better still, if canvas can be had, to tack it on the joints, and paper the whole partition on both sides with old newspapers.

Window-Frames. — There is a difficulty peculiar to log-houses in the putting in of windows, and it is this: In order to make a place for the window, it is necessary to cut through the logs, and consequently to weaken the wall. The part cut away must, therefore, be supplied with a good, stout frame capable of keeping these logs in place securely, and for that purpose it ought to be made to fill the thickness of the wall. This frame should be composed of two cheeks, say a foot wide and an inch and a half thick, stoutly nailed, above and below, to two cross-pieces, the upper one, or lintel, longer than the width of the cheeks apart, say two inches on each side, and two inches thick. The under one, or sill-piece, to be the same length of the lintel, but to be two inches thicker and four inches broader, so as to project that much beyond the outside face of the frame. On the cheeks, lintel, and sill of this frame narrow slats three-quarters of an inch thick, and say one and a half or two inches wide, are to be nailed within a couple of inches of the outer face of the frame; and against these the casement will shut. It will be hung on one side with leather hinges, which may be cut out of old boot-tops, and be secured at the other with a turn-button screwed on to the casement, and closing into a rabbet in the cheek, or jamb, of the frame. This casement, or window-sash, may be put together by any one in a rough way; but it would of course be better to buy such things in the neighboring town, where they are always to be had cheap.

However, as this book is intended for the use of those who possess little or no knowledge of the cunning of carpentry, we will here present the simplest mode of making a sash or casement:—First, you must be entirely governed by the size of your lights or panes of glass. These should be either 8 by 10, or 10 by 12, and supposing the former size to be the one chosen, and that you want to have but four lights of glass in each window, you will lay them out on a table, a board, or any flat surface, so as to form a square, leaving an inch space between them, and observing to keep the ten inches lengthways, to form the height of the intended window and the eight inches crossways to form its width, you proceed to measure for your frame. (Fig. 7.) For the outside have two pieces of pine, two inches wide by one inch thick, and twenty-four inches or 2·0 long. Have two more of the same dimensions, and twenty inches, or 1·8 long. Halve the ends of these down on each other, as shown (Fig. 8), and when you have fitted them together, so that all the surfaces of each will be perfectly flush or even with the other, then divide the distance between the lengths of all the sides into halves, and make a triangular notch at each of these divisions, one inch wide and half an inch deep. Next, put your frame together, as shown at No. 2 on the drawing, and secure it at the four corners with wooden pins or sprig nails. Now plane up a stick one inch wide, one inch thick, and three feet and one inch long. Cut this with a saw accurately into two lengths, respectively one foot eight inches and one foot five inches. Again, cut the 1·5 into two equal lengths of eight and a half inches each. Now cut all the ends into angles to fit the notches in the frame, and put the whole together, as shown in No. 3 on drawing. Tack down all around the frame laths three-eighths of an inch thick and one and

Fig. 7.

Fig. 8.

three-fourth inches wide, leaving one-fourth of an inch on the inside for the glass to bed upon. Tack down similar laths, only half an inch wide, on the rails, and the whole is ready for the lights of glass, which may then be set in and puttied. (Fig. 9.) The putty is made with whiting, worked up well with sweet or linseed oil with a knife. This completes your window, which is then set in its frame in the wall pressed from the inside against slats, and temporarily secured with inner slats, so as to permit of its being taken out at any time. Or, it may be hinged with iron or leather hinges; or be made to turn on pivots, either lengthways or sideways. But, whatever way is determined on, it is very necessary that the piercing wind and penetrating snow should be carefully guarded against.

FIG. 9.

In the winter season, in Alaska and other northern climates, it would be advisable to use double windows, as is the practice in the lower part of the Canadian Dominion. This effectually keeps in the heat, and keeps out the cold of the most piercing weather. In milder weather the second sash can be laid aside.

Doors.—The making of a door, such as will answer the purpose of a log-house, or humble dwelling of any description, is a matter so easy to be understood, that it requires no illustration to show the mode of construction. A door may be from six feet to six feet six inches in height, and from two feet six inches to three feet in width, according to the dimensions of the frame it has to fill. The simplest method of putting it together, is: Lay out on the floor, or on even ground, as many boards as will form the width of the intended door, and all to be of sufficient length. It matters not whether some are longer than others. Plane the edges smooth, so that the boards shall lie close together. Next, take a similar board to these, and saw it in two. Now mark off the height and breadth of the door according to the door-frame, which is already up, and putting the boards as close together as possible, nail the two short boards across the width at about six inches from the top and from the bottom. This done, square over the long boards, top and bottom; saw off the ledge boards (just nailed down), and plane the edges. At the middle of the height nail down a third ledge across the width. If this latter is wider than the other two, so much the better. Lastly, turn over the door, face upwards, and tack on laths over the joints. This will make the door weather-tight, and more than answer the purpose of tongued and grooved boards.

Butt, or strap hinges are so cheap, that a door can be readily hung with either. But in case even these are not on hand, there are two ways of hanging the door: one is by cutting up an old boot or shoe, and making substitutes for the iron article; the other is to bore a hole in the bottom and top of the door, at the end where it is to be hung, and fit tight into each of these holes a round oak pivot. Bore corresponding holes in the cap and sill, and cut out a square of the sill in which the hole is, and fit it on to the pivot, returning this square piece to its proper place, and nailing it there when the door is hung. Doors turning on pivots answer very well, and cost nothing, save the trouble of making the pivots and boring the holes. They can be made to fit close, and are both durable and strong.

If the door is not thick enough, nail on cleats above and below.

The latch and string is perhaps the simplest mode of temporary fastening. It is nothing more than a stick of hard wood ten inches or a foot long, bored at the inner end, and made to turn freely on a round pivot or screw the outer end projecting beyond the edge of the door and sustained by a cleft cleat nailed to the door; in the door-jamb a notch is made to receive the latch when the door is shut. The outer or entering part of this notch should be rounded, so that the latch shall glide easily over and drop into its place. A hole is bored in the door a few inches above the latch, through which a string can work freely, having one end attached to the latch, and knotted at the other end which hangs on the outside of the door.

There are other simple modes of fastening or latching doors; and the ingenuity of the handy man will, no doubt, suggest several, but none more easily made than this.

An oak or ash bar, three inches square, will be sufficient for permanent security; the ends to rest in mortises in the jambs, or side-posts, of the door-frame. The upper part of one of these mortises to be cut on a slope, so as to allow the end of the bar to slide into it, while the other end enters a square mortise in the opposite door-jamb.

The front door is sometimes divided into an upper and a lower section; each half being hung on its own hinges, and having its own fastenings. For keeping animals out, and children in, this half-door system is a good one, and frequently answers an excellent purpose.

The back door will be the counterpart of the front, and therefore the mode of making it need not here be repeated.

There is another and a very strong manner of door-making, which will always serve a good purpose when sound boards of sufficient length for the other mode are not available. It is, to saw all the boards into three-foot lengths, and cleat them together by three strips, three or four inches wide, an inch and a half thick, and the length of the door.

One of these will be nailed down the middle of the door, and the other two on either side next to the edge.

There is a method of making doors by placing the boards diagonally; but the necessary waste of stuff in cutting for this purpose would more than outbalance the advantage of bracing strength which it is supposed to possess.

LOG-CABIN WITH ONE ROOM.

THE dwelling here illustrated is of the most primitive and unpretending construction, and one which can be generally seen in the bush. But as the aim of this little work is to meet the wants of a very large class who have not even the knowledge which the building of such humble tenements requires, and who possibly may

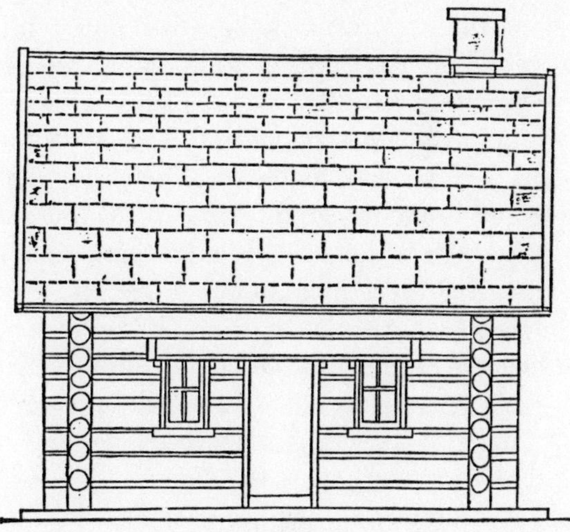

FIG. 10.

not meet with settlers more experienced in such undertakings than themselves, the timely hints here given may not fail of proving useful.

For a young couple with a small family, this cabin,

carefully built, will prove quite sufficient for present wants, and can be easily added to on the chimney end, so as to give any required accommodation without at all interfering with the first erection presented in this drawing.

The logs to form the walls should be carefully selected, sound and straight, and as uniform in thickness as it is possible to get them. They should be cut down so as to fall towards the site of the intended building, and thus save labor in carrying them to their location; and they should average at least two feet more in length than the walls call for. This will ensure the quoins being free from any liability to slip, and permit the cutting off of the ends without injury to the walls.

Supposing the illustration to be a fitting design for the proposed cabin, the course to be followed in the construction of it is as follows: Cut down all the trees of every sort standing on the chosen site, and level and beat down the ground. Now, roll together all the trunks (with their roots and branches cut off), and place them side by side, sinking those which are thicker than others down in the ground by digging under them, if necessary. Make the layer of trunks as level throughout as possible. This done and the place of the cabin marked out, cut down two trees, giving a fair trunk, of twenty-two feet long; and two trees of sixteen feet long, and whichever has to lie crossways on the layer or platform just formed may be sixteen or eighteen inches in diameter. The bed, or under side of these two sleepers, must be axed square, so as to ensure their firmness of position, and the upper side must be deeply notched, with the axe, at the distance, from centre to centre, figured on the plan shown in the next illustration. These two sleepers may now be deposited in their intended places, and the two other trunks laid across them, within a foot or so of their ends. Chalk off on these the places for the notches, and cut them out with the axe to correspond exactly with the notches already made in the sleepers, and roll these two trunks into their places. The notches should not be less in depth than one-third the thickness of the logs; which, fitting down into the deep notches in the sleepers, will leave but a small interval between the platform and these last logs; which interval should be closely wedged up, and have every chink filled tightly and permanently. This filling is of great consequence; as the warmth of the house in winter, as well as its ability to keep out vermin, will greatly depend on this first course. Now, mark on the front and rear logs, where the doors are to be located; chalk off the width of the respective intended openings, clear of the door and window-frames. Back from these chalk-lines, a few inches, place saddle-pieces made, as shown, next to elevation of cabin; and having placed a pair of these under the front and rear logs at the places named, saw these logs through for the width of the doors only, and so proceed placing logs notched down upon each other, and fixing the saddle-pieces for three courses, or until the course of the window-sills is reached. Then saw off the front and rear logs near to the location of the saddle-pieces, thus leaving sufficient opening for the windows, as well as the doors at front and rear. Also place saddle-pieces and cut the logs for the window-opening in the end opposite the chimney. The same course must be pursued in the making an opening for the chimney.

These saddle-pieces should not be less than six inches thick, and they should also be nearly, if not quite, the width of the wall's thickness. If they are got out of some tough wood, it will be so much the better. The notch, or cut, in them should partake more of the V than the round form, so as to prevent any possible slipping round of the logs, an accident which would be sure to wind the walls out of the perpendicular and weaken the whole construction.

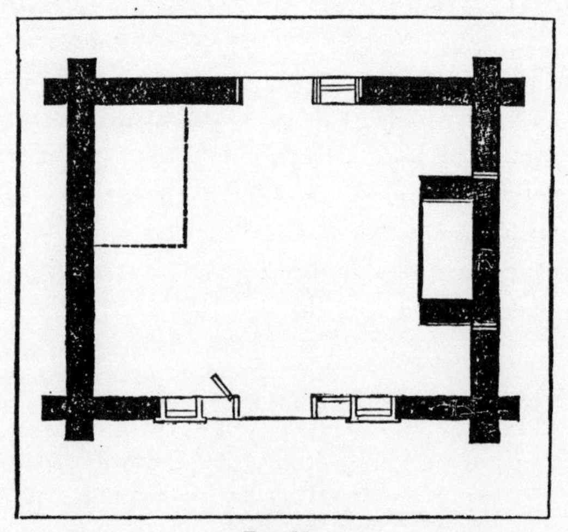

FIG. 11.

THE PLAN.

The length within the walls is eighteen feet by twelve; a space just affording sufficient room to do household work, to eat and sleep in. In many parts of the country the length of such a cabin is limited to fifteen feet; but, taking out the space occupied by the bedstead, and that taken up by the projection of the chimney-breast into the apartment, such length is barely sufficient for the purpose. Where the material actually costs nothing, save the labor of cutting and handling it; and as the same amount of cutting is required whether the logs be short or long, there is surely no object in limiting the dimensions to an inconvenient degree. The front and back doors are each two feet ten inches wide, and six feet high, having an angular hood over them projecting three feet, and connected with the shed-hoods over the windows, which project eighteen inches — all being supported by simple brackets, which, instead of being boards, may be tastily formed of forked boughs.

Rustic seats may be easily constructed of slabs. In summer they will prove most desirable as a place of out-door enjoyment; and in winter they will guard the base of the walls from snow and rain.

The chimney-breast is six feet wide. The fireplace

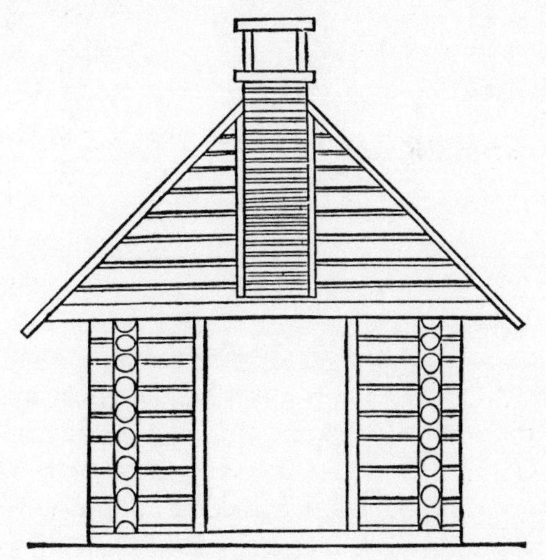

Fig. 12.

is four feet in the clear, and two feet six inches deep; thus affording ample room for large logs, such as our dwellers in the bush are in the habit of using; the custom being to keep the one fire burning all the winter through.

The mode of building the chimney of earth has already been described; but, if stones are at all to be had, it would be far more advisable to use them for the purpose, cementing them with mud-mortar, if lime is out of reach. The weather-side should be protected by boards, until such time as lime could be procured to make good mortar and point up the joints (previously raking out the mud).

The method of building the chimney for this plan is shown in the preceding drawing (Fig. 12), which represents the end of the cabin at which the chimney is located; and the construction is of narrow boards or slabs laid alternately, with a layer of mud-mortar two or three inches thick between every two of these, securing the whole with upright strips, or boards, at the angles nailed to every layer of boards, as shown. The chimney-shaft is carried up in the same manner.

The windows are each eighteen by thirty-four inches, having six 8 by 10 lights of glass. These window-sashes are to be had in every village and hamlet throughout the country ready-made and glazed. But should it so happen that they are not convenient, the simplest mode of making them will be found in another part of this book clearly and explicitly illustrated.

The hollow space between each window and the door, seen on plan, can be fitted up as presses, having doors, as shown.

It will be seen that the logs are cut short for the front wall to admit of this whole frame containing windows, door, and presses. But the logs above their heads should be the whole length.

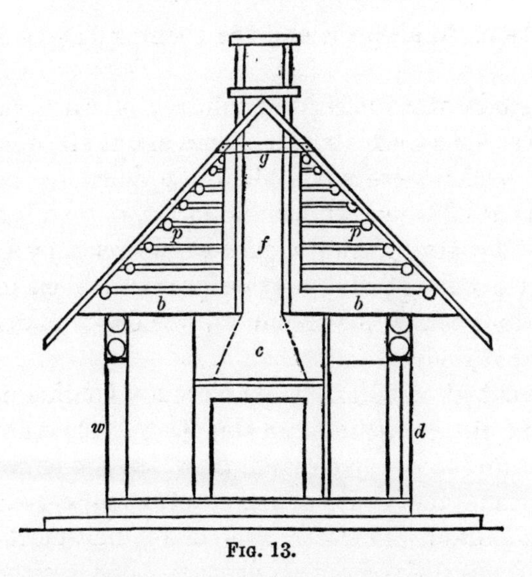

FIG. 13.

SECTION ACROSS INTERIOR.

In this illustration is shown the form of the chimney and its flue; the rafters of the roof and the mode of binding them with the collars; the purlins on which they rest, and the projection of the porch-roof, &c.

The floor may be formed out of rough boards laid on the platform of trunks, or logs, first spoken of. Previous to laying this boarded flooring, the spaces between the logs should be fairly filled up with gravel, dry sand, or pounded clay, until all is even, and the boards lie on a level bed. They may then be nailed down to the trunks wherever they touch them. Care must be taken not to have the boarding nearer than three feet of the fireplace, as shown by the dotted lines on plan. The ends of the boards should rest on pieces notched down on the logs of the platform, and should be carefully nailed down upon them.

If no flooring-boards, or rough boards of any description, can be had, the platform should be covered over, from two to three inches deep, with —

Three parts of washed sand;
Two parts of wood ashes; and
One part of clay;

the mass well worked together and wet sufficiently to make a plaster. This can be levelled off on top with a smooth, flat board of twelve or fourteen inches square, to which a long, oblique handle is secured by a cleat. Previous to finishing the surface, however, it will be advisable to roll the whole floor; and for this purpose a large, heavy log of bass-wood, about two feet long, and perfectly clean and even on the surface, should be procured. Into the centres of its ends should be inserted pivots of hard wood, and a square frame two by three inches should be then nailed together, having projecting cheeks with holes to allow the pivot to turn in. With this roller the whole floor should be gone over, until a firm bed is formed for the finishing coat, or smoothing operation, which is then commenced.

Before the floor is commenced, the covering should be put upon the roof; and, in this case, that covering is proposed to be of carefully selected smooth bark cut from ash or elm, beech or maple. This bark should be evenly cut in two-foot lengths and sun-dried. It is nailed on to the roof-boarding (or, in case there is no boarding, to the rafters), commencing at the bottom (the same as shingling), and giving each course a lap or weather-cover of ten or twelve inches. A double layer of these bark shingles would be ample security against the weather, even without any boarding. The eaves or edges of the gables should have tree-slices or slabs nailed down upon them, to prevent the wind raising them; and the ridge, or top of the roof, should be securely covered with sections of tree-trunks cut hollow in a Λ form to suit the angle of the roof they are to straddle, as shown.

b b is one of the beams which sustain the garret floor, or loft. These should be twelve inches apart, and have the upper surface axed to receive the boarding. A firm floor to the loft is a comfort which it only takes a few more tree-trunks to secure. Peel off the bark from these beams, and tack on slats wherever openings or seams occur in the boarding of the loft floor. Care in this matter will ensure cleanliness below.

c, The chimney, showing how to gather it in so as to connect with the flue *f*. This will ensure a fair draft. The inner face of this chimney and flue should be coated with a plaster of clay, wood-ashes, and fresh cow-manure, well worked together and smoothly finished. This coating gets very hard and is fire-proof.

d, Section through the doorway, showing one of the jambs, the cap, and sill. *w*, A section through the window, showing the pent, or hood, over it.

f, The chimney-flue, showing how it is formed and supported.

g, The collar-beam, strongly nailed to the rafter.

p, The purlins which sustain the rafters. These could be dispensed with; but they strengthen the roof very much, and cost nothing but the cutting. Besides, where they are used, the rafters may be comparatively slight poles, the ends of them cut smooth, and showing under the eaves.

The gables are formed of slabs or slices of trees securely nailed to the rafters and to upright pieces on each side of the chimney-flue. The bark may be left on, and the flat side turned in, and the seams slatted (as shown in section).

The chimney is covered with a box, open on two sides (*E* and *W*). But the covering described in another part of this book is far preferable, though somewhat more troublesome to make. However, the security from down-puffs of smoke, as well as the insurance of perfect draft, are advantages which it would be well to spend a little more time and labor in securing.

In the fitting-up of the interior, the recesses at either side of the chimney might be shelved, or inclosed as closets. The lower part of one being a flour, and of the other a meal bin.

LOG-CABIN WITH THREE ROOMS.

THIS illustrates a larger construction than the preceding, being adapted to the requirements of a family of four or six members. Its interior dimensions

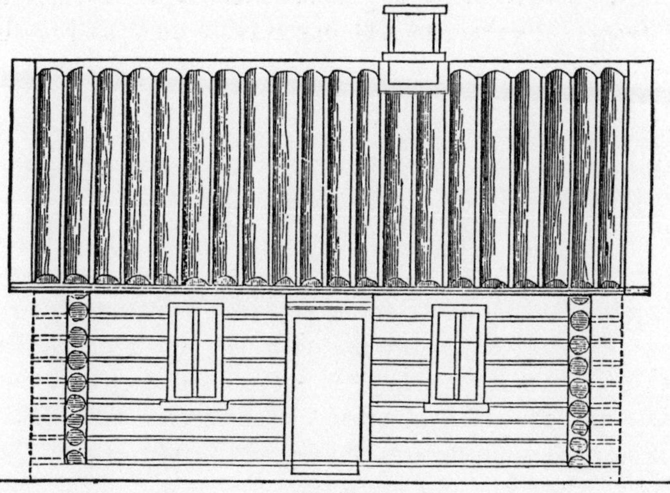

Fig. 14.

are twenty-two by twelve feet, and its height is eight feet up to the bottom of the flooring-joists of loft. There is a platform projecting front and rear, the making of which is described in the preceding pages. But this platform may also be made of earth dug out of a cellar, which may be under the living-room. It may be bounded or confined by trunks of trees, or it may terminate in a slope; whichever is deemed best for permanence or convenience. In any case, the raising of a dwelling on a platform contributes much to its dryness, healthfulness, and appearance, and is a mode which should always be adopted.

It will be observed that the structure now under consideration, although of fair dimensions, does not really call for long logs. The two ends requiring the longest; and these are left whole, because they act as stays to the front and rear walls, and, not having windows in them, render the house more comfortable in winter. The walls may be built in a similar manner to those already described.

The roof may be covered with slabs, as shown, and make a good job; tacking on weatherslips over their joinings.

It is customary to have the smallest size glass, and but four lights in each of the four windows in those pioneer structures. At one side of the front and rear door-cases there is a window, the frame of which is attached to the jamb. Half-way between the other jamb and the end wall (front and rear) there is another four-light window. The consequence is, insufficient light and air in these dwellings; compelling the inmates to open one or both doors, when it would be more comfortable to keep them shut. Good-sized windows do not materially reduce the strength of a log-house, if the logs are only judiciously secured at the openings, either pinned or saddled together, and that the jambs are sufficiently thick and strong to brace and support them.

THE PLAN.

The chief feature in this arrangement is the locating the chimney in a central position, instead of at the end wall. This gives the advantages of retaining all the heat of the fire within the dwelling, and warming the

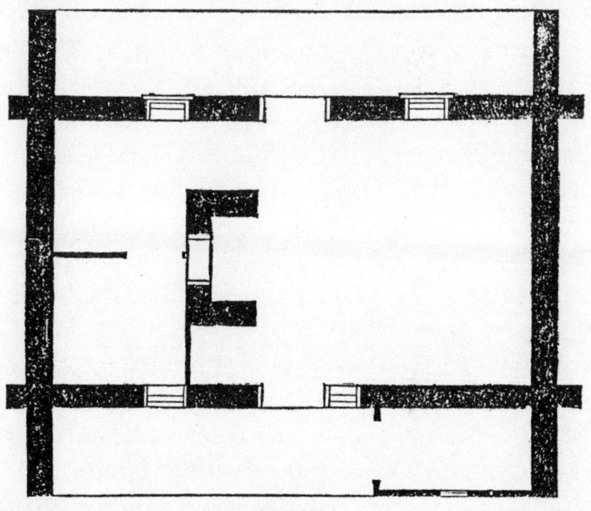

FIG. 15.

two sleeping-rooms. This latter advantage can be even better secured by cutting a round hole in each of these rooms through the back of the chimney, and inserting flush with the fire back a square plate of sheet-iron. It will be necessary, however, to protect these openings by a *thimble* of terra-cotta or sheet-iron.

In order to secure as much light as possible, sashes have been inserted in the front and rear doors, which are hinged, and thus admit air without opening the door to poultry, &c.

The two sleeping-rooms are small, but snug; and by

having the one open into the other, one of the deep recesses at the side of the fire can be used as a pantry for the kitchen. To reach the loft, there should be a manhole, sixteen by eighteen inches, cut in the ceiling ; the ladder belonging to which might be located against the side of the chimney, and be also available for descending into the cellar (if there is one under the house). Of cellars, and their proper location, we speak elsewhere.

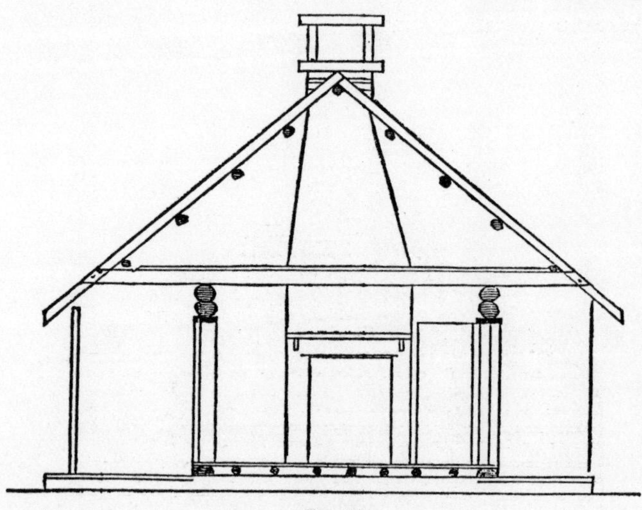

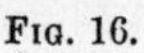

Fig. 16.

THE SECTION.

The throating of the chimney is quite different in this arrangement from that in the others. The contraction is more gradual, and the consequent shortness of the flue ceases to be an obstacle to the draft, owing to the application of the ventilating doors to the chimney-cap, every two of which being connected by a swivelled rod, resist the wind, and at the same time let off the smoke, so that there must be an uninterrupted draft.

The mode of flooring, as well as of forming the platform, can be easily seen and understood.

The roofing is made with three pair of purlins to support the slabs, without under-boarding.

The gables are covered with vertical boarding nailed to framing.

The eaves-gutters are sustained by the ends of the projecting logs and by two brackets at the sides of each door-frame (front and rear).

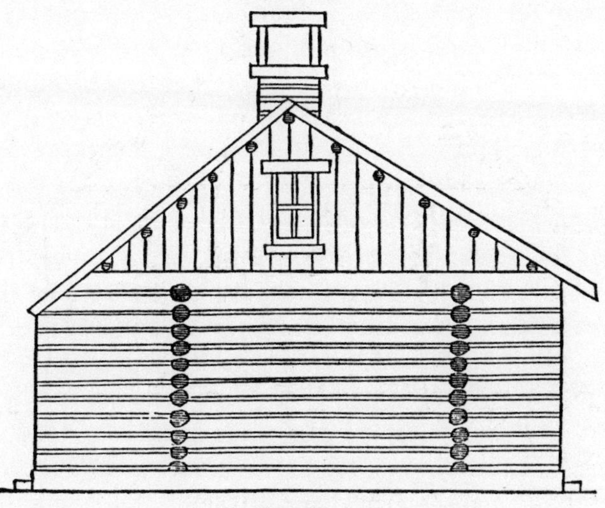

Fig. 17.

END ELEVATION.

This elevation differs from that of the preceding illustration in the mode of enclosing the gable, which is by vertical boarding nailed on to the before-named framing.

There is a window shown in this, and there is to be also one in the other gable for light and ventilation.

although it is not proposed to use this last for anything save storage of corn, &c., as well as for such loose matters as might be in the way down-stairs.

STABLE AND COW-HOUSE.

These very necessary appendages of a settler's home might in this, as in the preceding plan, be well located at either of these ends, having a lean-to roof against the wall of the cabin.

This building might be eight feet wide, and extend seventeen feet; that is, the clear width of the cabin, from out to out of the projecting logs. It might be partitioned midway, and have one-half for a cow-house, and the other half for a stable.

The walls could be made of slabs set up vertically, nailed to a stout sill at the bottom, and a stout-framed cap at the top. The roof might also be of slabs; and if two sets of slabs could be used for both purposes, it would add much to the comfort of the animals as well as to the permanency of the structure. A part of the loft of the cabin might be partitioned off (that over the living-room would be the more desirable) to temporarily answer the purpose of a hay-loft, and the stable being built at that end, the place where the window is might be made into a doorway through which to receive and to give out the hay.

The feeding-stalls should be at either end of the stables, and the two doors might be together at the centre, where the partition comes. Narrow windows should be provided in both front and end walls; for animals, especially horses, need light.

The floor of this stable ought to be paved, and have a gutter, or drain, to empty into a cesspool, or sunk barrel, at a little distance from it, and convenient to the dung-pit.

At the end of the house opposite to this stable and cow-house, there should be an enclosed shed for tools and farming implements, hen-roosts, &c.

The piggery should be kept away from the house; but be well sheltered, and have water convenient.

HEWED LOG-HOUSE.

THE difficulty in obtaining brick or stone, and the desire to still retain the solid comfort of the log-house without its primitive uncouthness, often prove sufficient arguments with the thriving settler to improve his condition by the erection of a farm-house of hewed logs; that is, of logs not alone stripped of the bark and trimmed, but wrought square with the axe.

Too much care cannot be given to the making the

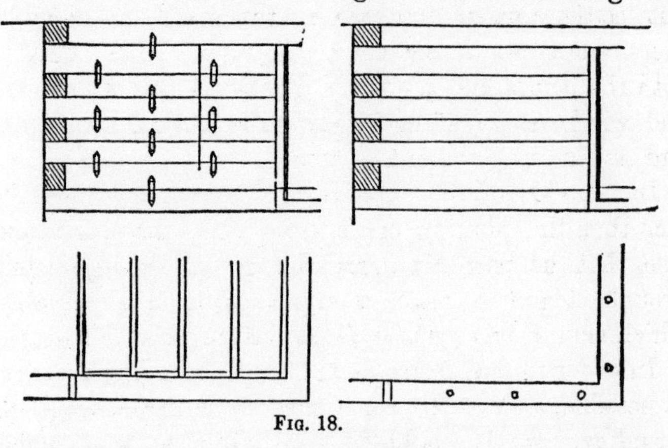

FIG. 18.

beds of the logs to form the walls as even as possible; and with this view, it will also be very necessary to choose the straightest and finest-grained logs, as the least wind in the thread of the grain will give a twist to the bed of the log which cannot be remedied, and would inevitably spoil the wall, if built into it.

The *section* on the left, in the accompanying engraving, represents the centre or heart of the wall; and it will be observed that the quoins, or corners, are not halved or notched on each other, but are laid together, being held in their places by pinning the several logs together, one to another, throughout their length. The door- and window-frames being made of two-inch plank, and stoutly pinned to the ends of the logs they are located against, will give additional strength to this construction.

The plan underneath this *section* marks the position of the pin-holes.

If, however, it be deemed too laborious to make and insert so many pins as this plan calls for, quite sufficient strength and security will be obtained by only driving pins at the corners, and pinning the heads and sides of the headers, or alternate logs, together. This, together with the before-mentioned pinning of the jambs of doors and windows, would make the walls sufficiently secure, and save a large amount of labor.

In the view of the exterior, also presented, it will be seen that the pinning or fastening does not show; and also, that the partially rounded edges of the logs, when brought together, make a regular system of *cinctures* which should be carefully filled and plastered flush up.

In order to make this mode of wall-building as warm as possible, and at the same time to secure it from internal *dry rot,* it would be very judicious to spread a layer of air-slaked lime between each pair of logs, previous to driving them together. The plastering of the joints would keep these layers of lime in place.

The wall-finish on the inside should have thin furring slips on which to nail the lathing, so as to secure perfect evenness of surface. But a fair job of plastering could be done on these inside walls by hacking up the surface of the logs and spreading a thick brown coat of lime and hair mortar, on which the skim coat could be laid.

The ends of the logs forming the corners should be sawed square off, and be rubbed with a clean stone until quite smooth.

The *plan*, under the view of the *exterior*, shows how the flooring-joists are supported. Instead of chiselling out holes in the logs' side to receive the joists, the mode here suggested is to pin on pieces to the cheeks of the logs, say three inches thick and five or six inches deep. On this tier, or ledge, the joists can be notched down, or let to stand without notching. But in applying this plan to a second story-floor, it would be very necessary they should be notched down, so as to bring the ceiling line even with the bottom of this ledge.

SAWED LOG-HOUSES.

If, as in many cases, a saw-mill is within reasonable distance, these logs, being cut into eight-inch, six-inch, four-inch, or even two-inch thicknesses, will, by being pinned together as just described in the preceding mode, make a comfortable and cheap house.

The *two-inch log-house* would be greatly strengthened, as well as made much more comfortable, by covering the outside with tongued and grooved flooring-boards, set vertically and slatted over the joints.

On the inside, these sawed log-houses can have the walls papered; previously covering the joints with strips of brown paper, pasted on in sufficient widths. They can also be furred, lathed, and plastered with one brown coat, and then be papered; or they can have two coats of plaster floated.

UPRIGHT, OR VERTICAL LOG-HOUSE.

WHERE taste can be combined with utility, this mode of constructing affords a very good chance for a display of truly rustic design. Strength, too, can be obtained by carefully using the sills and caps of windows and doors as binders.

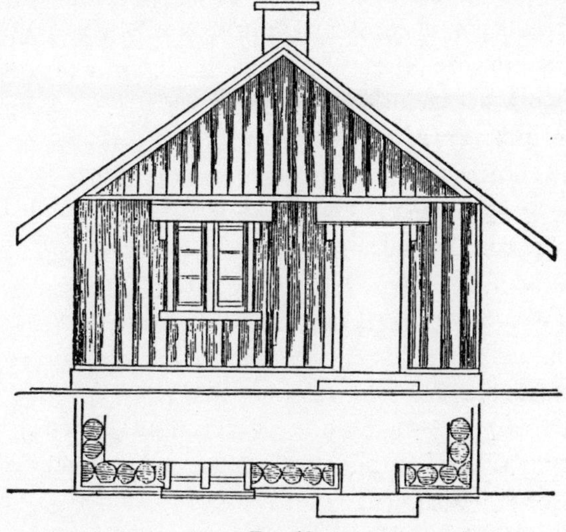

FIG. 19.

The sill of the house, and also the cap, should be strongly halved and pinned at their respective angles; for on these two mainly depends the stability of the upright-log manner of building.

An inspection of the PLAN beneath the WALL ELEVATION will give an idea of the construction.

When the sill of the house is framed, set permanently in its place, and firmly pinned together at the angles, the posts, or logs, which are to occupy the corners, and those which are to be next to the door and window frames, are to have auger-holes in their centres at both ends of each of them. The sill is also to have auger-holes at these required places, as shown on the plan; and into these latter are to be tightly fitted oak pins an inch and a half in diameter. Similar oak pins are to be fixed in the upper ends of the logs in the holes prepared for them. These pinned logs will now be set upright, their lower ends, or feet, shut perfectly down on the pins standing in the sills to receive them, and temporary strips will be tacked along each wall to keep them in place. The other, or intermediate logs without pins, will now be set up until all the spaces are filled. The cap pieces, front and rear, will now be shut down tightly on the pins fixed in the before-mentioned posts or logs, and the angles halved on each other, bored with auger-holes, and driven down on their respective pins.

An inch-board, ten inches high, will now be nailed on, around the inside, to the base of the upright posts forming the walls; and a similar board, inch and a half thick, on the outside of the base of the walls, to form a plinth. The flooring-joists will next be set in place, bearing on the sill, which, for that purpose, will project on the inside three inches. The joists will abut against the base-board already mentioned, and when the flooring boards are laid, the walls will be immovable at their base.

The ceiling-joists will next be set in place, and nailed down, and the framing of the roof put together and covered in with boards, leaving an opening for the chimney shaft to come through. The roof may now be shingled and the chimney built, so as to afford an opportunity for the enclosure and completion of that part where the chimney rises, without making two jobs of it.

The window- and door-frames will next be inserted and secured. To save trouble, it is more desirable to have double windows, such as shown in the elevation, than to have separate windows spaced apart.

The partitions dividing the plan of the house may now be set up, and all the walls and ceilings lathed and plastered. One of the plans already given for log-cabins will answer, according to size of family.

The roof should extend at least two and a half feet beyond the walls; and if more than this, so much the better, for in the country there is nothing more acceptable than shade from the overpowering heat of the summer sun, especially on the prairies; and few things more conducive to healthfulness than the immunity from damp walls which is thus secured by the thorough shedding of the rain or snow.

If the roof is to have gables, it will be well to have the posts, or logs, of graduated lengths, to whatever slope or angle it is to be. But if it is to be a hip roof, the posts or logs forming the walls may all be of one length.

In order to effectually exclude all possible inroads of the weather, it would be judicious to tack in slips between every pair of posts all around the walls; and if these are white-pine slips, and the bark be left on the logs, the contrast of light and dark colors will afford a very bright effect.

The base course, or plinth, around the outside ought to have a sloping weather-board, indented so as to fit snug up to the logs, and projecting, say half an inch, beyond the edge of the base board or plinth.

SLAB-HOUSES.

ALTHOUGH this mode of construction is under much disfavor on account of the very general use of it by canal and railroad laborers, yet that very fact is a most conclusive proof of its merit as a cheap and ready way of building; for the class in question cannot

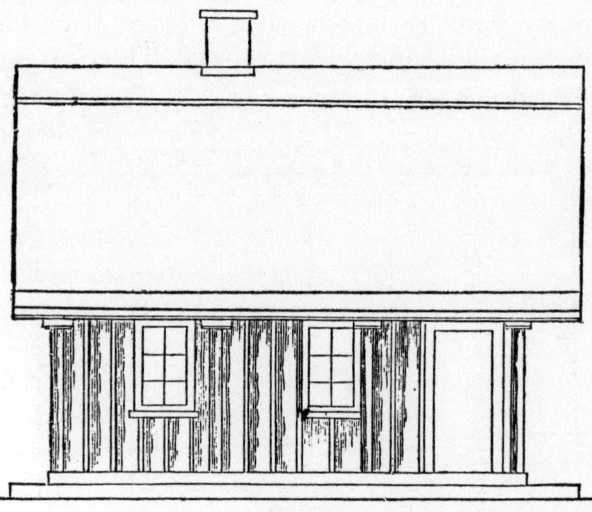

FIG. 20.

spare the time, nor afford the expense, of putting up any other dwellings. But it is not the mere thing of using slabs that is the objection to this construction, it is rather the invariable roughness and want of judgment with which houses are built of slabs. The aim of the present little treatise being to provide the emigrant and settler with the information necessary to enable him to put up a dwelling suitable to his means, and in as comfortable a style as his position will admit of.

In selecting the slabs for the walls, &c., it is very necessary that they should be of uniform width, and as free from large knots, knot-holes, and other defects, as possible. The edges should be axed even, so as to have them lie as close together as practicable. The bark had better be removed, as insects are apt to breed in it, and its presence will subject the slabs to premature decay. It, however, gives a very pretty rustic effect to the walls, and where slabs are used for other purposes than this of dwellings, the bark should be retained. The tree-trunks, forming the piazzas in front and rear, should have the bark left on.

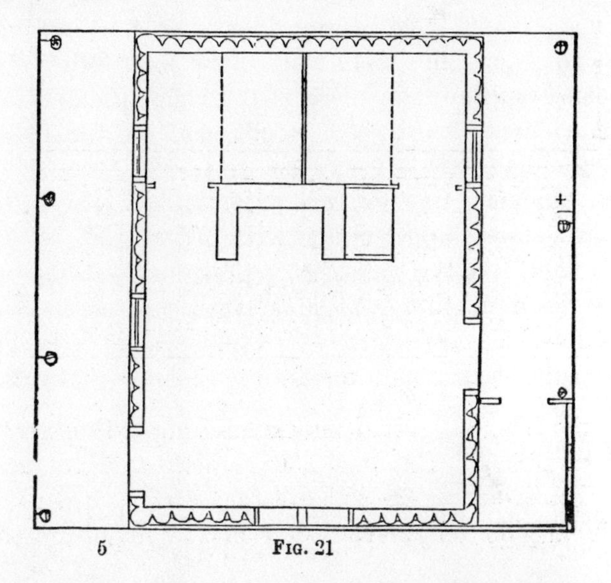

Fig. 21

THE PLAN.

The piazza, or platform, in the rear is not indicated on the plan, but may be seen on *Section*.

The choice of arrangement in these small dwellings is necessarily very limited; consisting, as they do generally, of but a living-room, or kitchen, and two small sleeping-rooms, with a loft overhead, which may be used for sleeping.

In this plan the living-room is too small to admit of space for a bed in it. It is a pleasant apartment, however, being well lighted and airy. The sleeping-rooms are unconnected; each having its separate entrance from the living-room, instead of having to pass through one to have access to the other.

The dimensions within the walls are fifteen by twenty feet, and the height from floor to ceiling is eight feet. There is a snug little cupboard, or pantry, at one side of the chimney; and there may be one at the other side, if deemed desirable, for there is the same space for one. If not so used, it will be a good place for the ladder under the man-hole to give access to the loft.

The chimney may be constructed in the way before described, unless brick or stone can be had.

The floor, like the walls, is composed of slabs laid and pinned down, crossways, on long trunk sleepers, flat side up, and bedded in dry clay or sand, which is packed between the sleeper, and up within an inch of the surface of the slabs, so as to have them firmly bedded before being pinned down.

A careful inspection of the plan will show the mode of construction, which is, to put together, strongly, and set up on edge a two-inch base-board (marked *a* on plan) all around the interior of the house. This base-board may be in width eight or ten inches. A similar construction may be adopted at eight feet above, or the ceiling line of the room; tacking it temporarily in its place at the four corners to posts, set up for that purpose, and firmly stayed on all sides.

The slabs, made even on their edges, may now be nailed on, above and below, to the cap and sill just described, leaving the corners, or quoins, to the last, when the posts and their supports at those points may be removed, and the remaining slabs nailed on in accordance with *the plan.*

The *platform*, front and rear, may be made of earth and stones, well packed or rammed between the sleepers outside of the walls, the surface being made of sand, ashes, and clay, puddled together and plastered on with floating boards.

A base-board is likewise to be nailed on the outside. It is to be the height of a step, and to be covered by the projecting door-sills.

The window and door-frames, to be made of two-inch plank, are to be the width of the thickness of the walls.

Slips are to be inserted, and nailed between the slabs on the outside, so as to make the joints perfectly weather-tight.

The flooring of the loft may be of slabs, evened on the edges, and made to fit close. They are to be nailed down to the flooring-joists, having previously been notched with the axe where they are to sit on those joists.

THE SECTION.

Here the construction is more distinctly seen. The section through the floor shows how small poles can be set between the slabs, so as to keep them from canting to one side or the other.

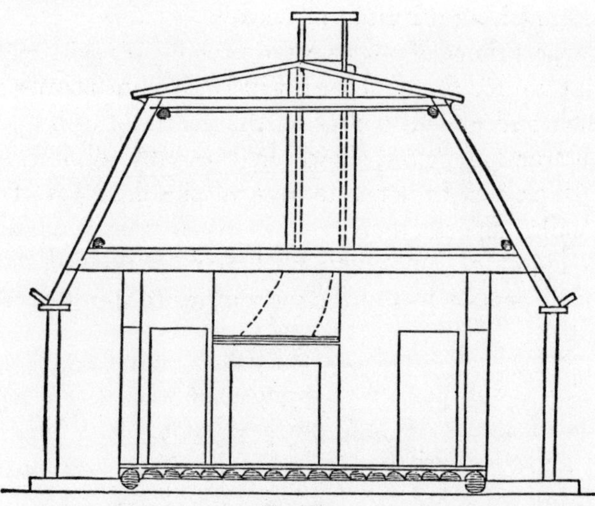

Fig. 22.

It will be observed also, that the platform trunks rest on slab sleepers, turned down flat, and sunk in the ground, their backs being on a line with the surface of the ground. This platform is the same, front and rear, and is intended to afford an opportunity hereafter to enclose, and make pantry and closets.

The ceiling-joists are made five feet longer than the width of the house, from out to out of wall; thus giving more space in the loft, as well as affording good shade with the aid of the extended rafters, to the front and rear of the house. These rafters, which are thirteen feet long, lean on two sets of purlins, and connect with the flatter section above. Their ends rest on slabs which are pinned down to the posts, and to which they are securely nailed.

The square window-frame is set diagonally, the slabs being easily cut for the opening, and the frame nailed on to them, making all firm and secure.

The partitions of the sleeping-rooms, as well as the sheeting up for the cupboard, will be of thin boards run together, and nailed to slips on the floor and ceiling.

The framing pieces, shown in the gable of loft, may be of thick boughs, axed to receive the flat faces of the slabs nailed against them.

The object of constructing a flatter termination to the roof is to present less unbroken surface to the weather.

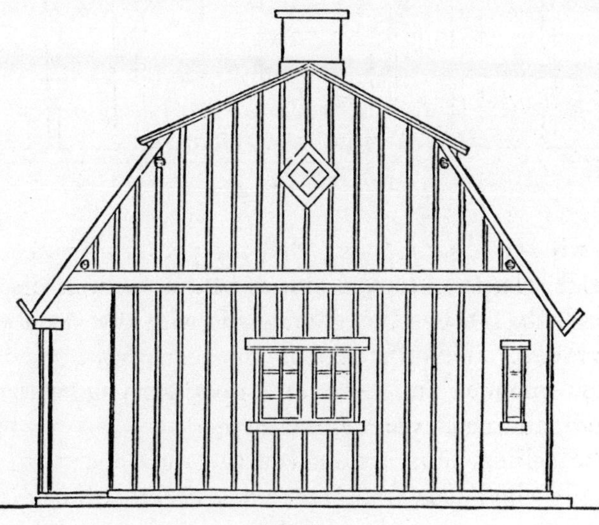

FIG. 23.

THE END VIEW.

This end is chosen for illustration, in preference to the other, because it has a double window in it, the accurate spacing of the slabs, in reference to which, being requisite.

This double window must be stoutly framed, and set securely between the slabs on either side of it, before the remaining slabs are put on, (excepting the corner ones.) There should be ledges to this frame, on the four sides, projecting on the inside, to which the slabs are to be nailed; that is, the two side slabs, and the top and bottom slabs.

The slips covering the joints of the slabs will not be used on the sides next the window-frame, as the frame itself will occupy their places.

The lower ends of the rafters will have a triangular piece, running the whole length of the house, nailed on to them, and this piece will rest securely on the reversed slabs, which form the caps of the columns, front and rear, and which slabs are pinned, with oak pins, down to the centre of each column.

The rafters will also be firmly pinned to the projecting ceiling joists, where shown on both *section* and *end view*.

It is never desirable to notch rafters, or indeed any timbers, as is so often done; all such cuts or notches reduce their strength just so much. Where it would be thought advisable to notch, or fit one piece to another, it would be far more judicious to nail on hard wood-blocks, on either side, one piece to hold the other.

PLANK HOUSES.

WHERE a saw-mill is in the immediate neighborhood, there are several ways of using boards or planks in the construction of houses. We here present parts of two modes of such construction. No. 1 shows the planks or boards (for it matters not what is their

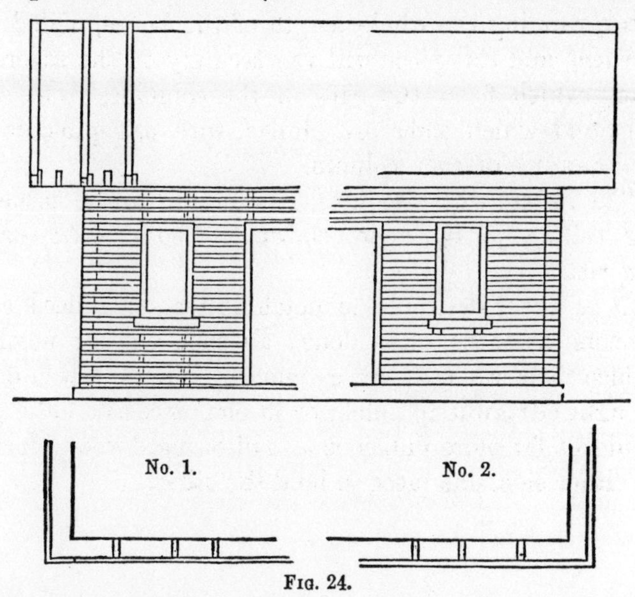

Fig. 24.

thickness) laid flat upon each other. Of course, there will be intermediate ones lying between those that lap at the angles. These may be pinned down to the others, although the tightness of the joinings at the angles would be quite sufficient to hold them in their places without any additional security.

No. 2. In this mode the intermediate planks, or boards, are left out, and in their places are blocks nailed down permanently, and vertically over each other, as distinctly shown in the elevation.

In case inch-boards are used, the ends will be too thin to nail the window- or door-frames to, and it would not be sufficient security to place them without some fastening, so that it will be necessary to cut in at two or three places, to admit of a block being inserted, to which the upright frames may be nailed. These blocks may be three inches wide, five inches long, and two inches deep; that is, each block will exactly fit a recess two inches by three, cut in three boards.

This, of course, applies as well to No. 1, if the boards used there are also inch-boards.

The objection to mode No. 1 is, that, the boards lying close, there is no chance for the drying out of any sap, which in all probability is in them. No. 2 has an advantage in affording a chance for plastering without lathing. In No. 1, this lathing privilege may be acquired by projecting every alternate plank, inward or outward, as the case may be; or in having every alternate plank an inch, or an inch and a half, narrower than the rest. This would leave the outside flush, and the inside prepared for plastering.

PLANK AND STUD HOUSES.

IN this illustration, economy of plank is the object. And for that purpose, square studs, the thickness of the wall in depth, are placed at the angles, and at every intermediate three feet. Each tier of plank is pinned

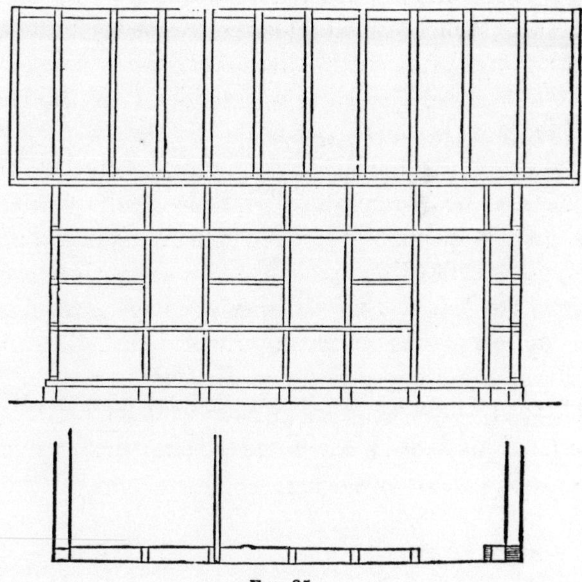

FIG. 25.

firmly down on these studs with oak pins. The feet of the studs have oak pins firmly fixed in them, and these pins are also tightly fixed in the plank, passing through it into the next stud beneath; and so on throughout the wall construction.

In providing for the window- and door-frames, it will be judicious to cut the studs in lengths, which will admit of the lintels coming in exactly under the tier of plank ranging on their height. And the same precaution may be exercised in the preparation to receive the sills of the window-frames.

The walls of this construction of house may be furred inside and out with narrow strips set vertically sixteen inches apart. On the inside they may be lathed for plastering, and on the outside they may be clap-boarded; or rough-boarded and shingled.

Where a brick kiln is adjacent, these hollow walls may be filled with dry brick to make them more comfortable.

Plastering on the outside is highly objectionable, inasmuch as the absorption of wet in rainy weather will cause the laths to swell, and in dry weather to shrink again, thus inevitably cracking the plaster. Clap-boarding is much better and safer as a covering against the direct action of the weather.

THE BALLOON-HOUSE.

THE name given to this mode of construction indicates its lightness and total want of any heavy element of solidity. Yet it undoubtedly possesses strength, and the facility with which it can be put together, gives it a peculiar claim on the man who desires to save time, labor, and money, in the erection of a ready home which possesses the capability of being rendered comfortable.

Frame together at the angles a stout sill, say four by six inches, which has been bored on the under side with an auger at six places (at the four corners and midway of the length). Set this sill on six stout cedar posts, driven four feet into the ground.

Next, nail up, at each of the four corners, a pair of boards abutting each other; and, to strengthen these, temporarily nail on the inner angle of each a pair of board-blocks at a couple of feet apart. This done, and the height of the house being decided on, chalk that height on the upper ends of these corner-boards just erected. Set a piece of scantling, three inches thick by four inches wide, along from corner to corner of end, and nail the upright boards to it. Do the same at the other end. Now connect these two end-pieces by similar pieces across the front and rear, halved down and spiked on the end-pieces at their angle of meeting. Proceed to board up the four sides, nailing them securely at bottom and top. Measure off for the location of doors and windows, and nail up boards where their frames are to be secured. When the flooring of the joists is all in place, and the boarding of the walls all up, then fit in and nail the window- and door-frames in their places.

Meantime the roof may be constructed. Run the ceiling joists out two feet beyond the walls, nailing them on to the front and rear pieces, and spike the rafters to the sides of them, at their ends, also spiking the rafters to one another at their tops. Or, better still, saw off and nail them to a ridge-board set on edge from gable to gable. This plan will secure the perfect uniformity of the roof throughout. Also, instead of spiking the lower ends of the rafters to the projecting ceiling-joists, nail flooring-boarding across these joists, out to their ends, and saw off the ends of the rafters, so as to fit down on this boarding, and spike them firmly down, through it, into the ceiling-joists. This plan will effectually enclose the eaves, without any further trouble. In the other case, the eaves will require to be boarded up under the ceiling-joists.

Saw off all the projecting ends of the upright boards of the walls, level with the upper edge of the ceiling-joists; and, where a joist comes, cut these boards accurately to fit against it. In order to make the construction perfectly weather-tight, close attention must be given to these matters, small in themselves, yet of infinite importance in the making a house comfortable.

Board over the roof; and afterwards saw out the hole for the chimney-flue.

If stoves are used, it is not necessary to build a chimney. Construct a flue resting on the ceiling-joists, or on a stout frame resting on the flooring-joists below, and have one or two stove-pipe holes, with thimbles in. If two, or even three, stove-pipes enter it, the size of the flue may be sixteen inches by twelve. If but one is to be provided for, eight inches by twelve will be sufficient. The frame on which this flue stands may be five or six feet high, and be enclosed so as to form a closet or locker. Cover all the external joints of the boarding with slips two inches wide and an inch and a quarter thick, planing off their outer corners. Cover the inner joints with rough slips, and these will answer for furring whereon to nail the lathing for plastering.

These slips on both sides of the inch-boarding tend to stiffen it very much. On the exterior they abut against a base-board below, and a fascia-board above.

The roof is usually shingled on rough boarding, and the exterior may be painted and sanded. The strips, or battens, as well as the trimmings around doors and windows, may be of a darker tint; or even be a direct contrast.

In order to make these balloon-houses warmer, they should be lined with thick brown paper on the inside of the boarding before the inside furring is nailed on.

A material called "building paper" is largely manufactured for this purpose, and may be had in any quantity in all the cities of the Union.

As this balloon-mode of construction is quite similar to that we have already treated of under the head of Slab-Building, the illustrations given for that will so nearly suit this as to leave no necessity for multiplying drawings.

It may be advisable, in this as in other cheap modes of construction, instead of building chimney-flues to use *terra-cotta* drain-pipes for that purpose. These can often be had in the nearest village, brought home, and put up, when bricks or bricklayers to build a flue may prove a serious, if not insurmountable want. Where it becomes necessary to make a continuation, two of these drain-pipes can be joined together by bass-wood splints secured with wire, and then coating this connection with mortar of wood-ashes, clay, and sand.

FRAME-HOUSE.

FOR those who desire more accommodation than is afforded by the foregoing arrangement of plan, we will proceed to lay other and more comprehensive designs before them.

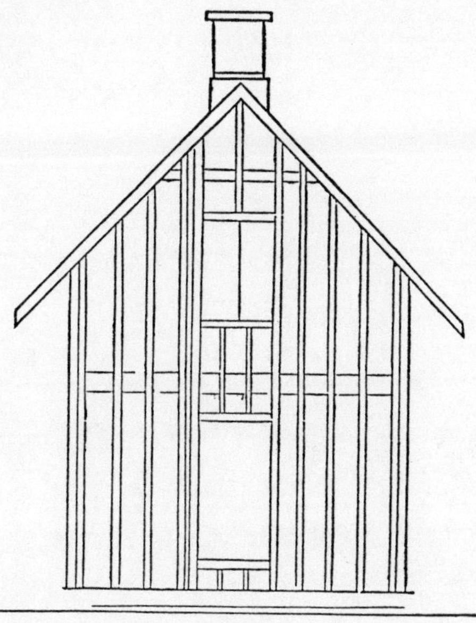

FIG. 26.

Wooden frame-houses are governed in their height of story by the length of corner-posts that can be obtained. Houses in all other material, such as stone, brick, concrete, béton, &c., are not thus limited, except in the matter of cost.

A wooden house (frame) of fifteen feet story-post will have its first story eight feet high, and six feet high on the second story, to the eaves; the remaining foot being for the thickness of floor. The ceiling of the second

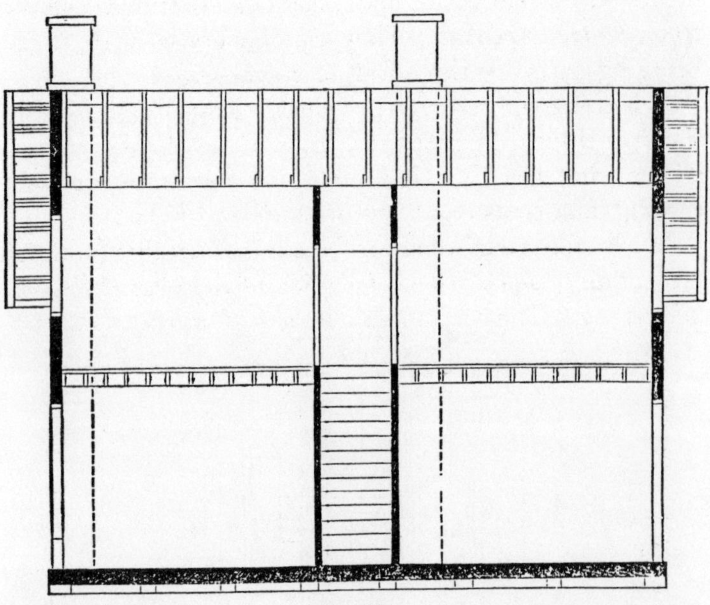

Fig. 27.

story will be the line of the collars of the roof, which in most cases will make this story, say, eight feet in the clear.

Or, the first story may be ten feet, which will leave the second story seven feet, — quite high enough for all purposes.

But in those economic houses, height of story is to be kept as low as consistent with appearance. Such rooms are easier heated in winter, and the amount of *stair-travelling* is not so great. Neither is there any object in having the corner-post more than fourteen feet; considerable saving being the consequence.

The studs in frame-houses of small dimensions may be proportionately small — in fact, one and a half by three inches will be sufficient, and placed apart sixteen inches from centre to centre. They may be braced with short bridgings at every three feet, or so, in height.

It is a source of comfort to sheet up on the studs with rough inch-boarding, before laying on the clap-boarding; as the wind has then less chance to penetrate the wall through any occasional split or chink in the latter.

Much more comfort might be attained in this (and, in fact, in any) house by the adoption of the plan of *double plastering*. That is: to lay on a stiff rough coat, and

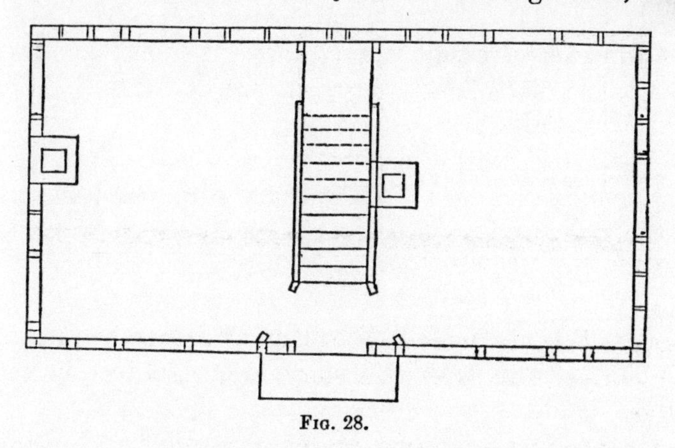

FIG. 28.

then nail on light furring on the lathing, and lath again. Now, put on two coats of plaster, and finish. This plan makes the wall in effect *double*, having an interspace between the two plasterings. It also stiffens the work to such a degree that it is secured against being easily broken through.

The ground-floor should have the cheeks of the joists provided with slats, or battens, nailed horizontally on to them, at about half their depth, and then between the joists should be sheeted with rough boards, resting on these slats. Finally, this sheeting, or slab-flooring, should be covered with some material, such as tan-bark, saw-dust, sand, &c. This advice has been already given, but it cannot be too much impressed on the mind of the man who is anxious to seize every available means of making his house comfortable.

We would also recommend the use of strips of felt to cover the joints of plank on the inside of the walls. If these strips are saturated in tar, it will add to their usefulness, not alone in keeping out the weather, but in effectually debarring all vermin of a refuge in the joints.

Brown paper would perhaps answer a similar purpose; but, being liable to split or tear, it is of course inferior to felt.

Too much pains cannot be taken to make the dwelling comfortable during the progress of its erection; for any opportunity neglected of adding these precautions just mentioned, as well as others which will suggest themselves, cannot be afterwards corrected, and will always prove a source of regret to the owner.

This plan, just presented for a frame-house, will equally suit the balloon-house previously spoken of.

CONSTRUCTION WITH EARTH.

WE have hitherto endeavored to point out and explain the various easy modes which the pioneer can choose amongst for a dwelling suitable to his position, his wants, and his means, where WOOD is the material most ready and most plentiful.

We are now to suppose his destiny is cast in a locality the reverse of this, and where, other materials being absent, he is driven to the necessity of using the earth on which he stands.

In this position, most generally, is the intending dweller on the wide prairie placed; and to him the knowledge of such construction as we are now about to describe is of sterling value.

PISÉ, OR RAMMED EARTH HOUSES.

To the French belongs the credit of the practical perfection to which this mode of construction, by them called *pisé*, has been carried. The principle on which it is founded is taught in the simple fact that a hard pathway can be soon made, even through newly ploughed ground, by the mere tracking of constant footsteps. This footpath, however, is hard only on the surface; yet it clearly suggests the idea that continuous layers, or strata, of such hardened surfaces must make a solid mass.

Thus we see that earth, by compression, is capable of forming a ready and reliable material for building. The military engineers of this country, no doubt borrowing the idea from the Romans of old, have long since built up earth forts, which have proved vastly preferable as modes of defence to any constructed of stone. But as such walls are sloping and necessarily very thick, they afford no hint, of any consequence, to our present purpose of using earth for the building of thin walls, such as are called for in the erection of cottages and other small dwellings, and which we will now describe.

Having cleared off the site of the intended habitation, and laid out the lines of the plan, dig the trenches for the foundations, two feet deep and eighteen inches wide; cutting similar trenches for the cross-walls, and one for the chimney, according to its dimensions, which will be, say six feet by three. Fill up these trenches with broken stone, gravel, and sand, or clay, all well packed and rammed. The rammer used for this purpose may be an arm of a tree, or any convenient pounder which is heavy enough.

Now, level the whole, and even it off, making the finished surface of sand and clay, mixed wet with a sprinkler of any kind, and smoothed or flattened with floats, such as described elsewhere. Level all around again. This process, which is performed with the instrument made for that purpose, and which is described on the page of "tools," is set upright with the long base resting on the wall, and when the plummet or weight, swinging at the end of the cord, rests fairly on the perpendicular line marked on the upright arm, the base is then perfectly level, as, of course, is the plane on which it rests. If it be not level, then slip bits of chip under the lower edge until it comes right. At the part where the chips were required, fill or level up with earth, until all is right, and slope off this additional earth to the points which required none. In this manner proceed the length of the level every time, until the walls are all levelled.

The plan must now be referred to, and the walls above the foundations laid out to their proper thickness, taking care that every corner is fairly squared before the work proceeds.

The manner of setting up the mould for making the walls is as shown in the accompanying diagram (Fig. 29). The mode of preparing this mould is: Cleat the two doors (front and rear) intended for the house. They will answer for the sides. Make one end (similarly) the breadth of the intended thickness of the wall. Now set on the foundation, at equal distances apart, three two-inch sleepers (Fig. 30), each thirty inches long and five inches wide, having two holes in each. Into these holes set the uprights (Fig. 31) (four feet long each), and after wedging them in, secure them at the tops with a strong cord or clothes-line, in the coil or double of which fix a short stout stick of ash, or oak, and twist it tight, letting the end of it bear against a pin secured in the side of each post.

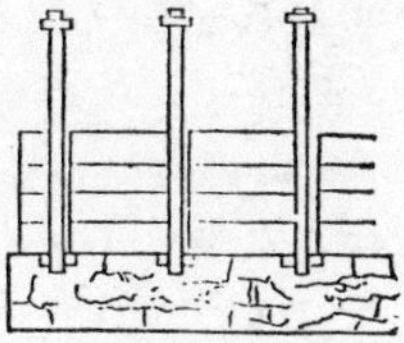

FIG. 29.

FIG. 30.

FIG. 31.

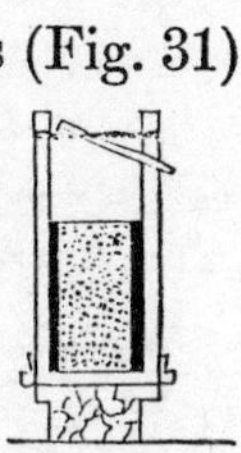

FIG. 32.

Next, slide the two doors, or sides, into their places, and insert the end, securing it with two nails passed loosely through two holes in either side of the post into two corresponding holes in the side-thickness of the end.

To keep the opposite end of the mould fair with this, fix a stick of the length required to stay the two sides in between, and all is ready.

The first setting up of the mould ought to be at one of the corners, and when the location of a door is reached, it would be well to have the door-frame already set up, so that it may be moulded up to. Projecting blocks should be nailed on the sides of the door-frame next to the mould, to insure it a good hold on the wall.

Previous to commencing the walls, it will be necessary to test the quality of the earth available for the work. Turn over the surface, and dig down until clay is reached, which will show itself in lumps on the spade. If, however, the ground dug gives up sandy or poor material, change the spot until the right place be found. Gravel mixed with clay is good; and even if big stones are found among the layers of clay, it will be so much the better. Gravel three parts and clay one part will make as good material for this purpose as need be had, as it contains the binding and strengthening elements. Fatty clay, or sandy soil, are both to be avoided. But if absolute necessity compels the use of both of these, they must be used with the greatest care. To mix the poor and the rich will require two men: one to throw up the poor in a heap, and the other to follow every three times with a shovelful of the rich. When the pile is complete, it is to be spread, chopped, and turned over, until all is thoroughly mixed. If it does not require this mixing, the clay is to be thrown up into a heap, and the stones and coarse gravel allowed to roll to the bottom. If any of the stones are larger than a hen-egg, they are to be broken small and mixed up with the clay and gravel.

Great care must be exercised in the working up of the

clay, that it be but slightly damped. The actual difference between the too dry and the slightly moist states requires much skill in the determination of the line of true distinction; for if the material be too wet, it will not bear to be rammed at all, and if it be too dry it will not adhere together as required.

But two or three inches should be the thickness of each layer to be tamped or rammed, and the men should frequently cross each other's work, and not consider a layer sufficiently rammed while the rammer leaves a track, (Fig. 33 shows the rammers.) For working around the edges of the mould, there is a distinct rammer (No. 2), which does such work with more nicety and certainty than rammer No. 1 can, on account of its peculiar shape preventing its working closer. The clay put into the mould must be perfectly free from all vegetable matter, such as roots, leaves, chips, twigs, hay, straw, &c., or anything whatsoever that can suffer any change, or is liable to decay, as the pressure of such would be sure to effect the future stability of the wall.

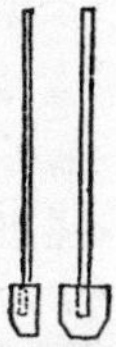

FIG. 33.

No more should be dug than can be used in a day. And if rain is expected, every precaution should be taken to cover up, and guard against injury to the walls. To this end, boards should be ready always.

As soon as the first layer is spread over the bottom of the mould, say three inches deep, the work is to commence by one of the men getting in and treading the whole well over evenly, taking care, in getting in and out, not to disturb the mould. Two men may now take rammers, and commence tamping or ramming, beginning at the sides, and then crossing in diagonal lines, going over the whole surface repeatedly, and perfecting each layer by edging around the mould with rammer No. 2. The next layer must not be commenced, until the one under it is so hard as to almost ring under the blows of the rammer. It will take three men to do the work: one to supply, and two to ram. Attention must be paid to the posts, that they do not lean out too much under the pressure of the filling and ramming of the mould, in which case it will be very necessary to tighten the rope or cord by giving the stick an additional wind around and again fixing it against the pin, (as shown in Fig. 32.)

As the filling approaches the top of the mould, stop within say six inches. Loosen the cord by removing the tightening-stick, then slipping it off of the top of the posts. Gently beat up the wedges at the bottom, so as not to jar the new-made section of wall, and carefully slide out the sides of the mould. Raise the posts out of their holes, and draw out, very gently, the sleepers from their places, and transfer them to the next section to be built, which will be a continuation of that just now uncovered. The pair of sleepers next to the succeeding block may be left in. Proceed as before. When the round of the walls has been completed, the second course is to be commenced half-way from the corner, so that the joints may be broken throughout.

It will be observed here, that the mould having but one end to it, the material necessarily has a slope at the other end; and it will be, therefore, desirable to set the mould for each succeeding section so that it will lap well on this slope, securing by this means the uniformity and required strength of the work. Where these laps occur, it will be requisite to tamp the fresh clay so thoroughly as to produce a perfect union with the block just completed.

Care must be taken to insert the window- and door-frames accurately in their proper places. On no account should furring-pieces be inserted in the walls to nail trimmings to; but when the skirting comes to be put down around each room, let it be nailed to the floor close up against the wall, and have a strip two inches square nailed against it, also to the floor, to secure it. Or, better still, nail this strip on to the bottom of the skirting before putting it down, and then nail the strip, with the skirting attached to the floor, up to the wall.

The cross-wall, or walls, will be worked up at the same time with the main walls, having a tie on them; that is, a cut, say of six inches, made with the spade in the side of the main walls, to allow of the cross-wall being moulded into them at its junction.

Where the chimney is located, the breast may be moulded solid, and after the height for the mantel-piece is reached, then a square block, say eight by ten inches, and as long as the chimney-breast is wide, is to be inserted, and worked over. The fireplace can then be cut out with the spade, and the flue behind the log be also moulded in by means of a square box of the necessary dimensions, made for the purpose. This box should have two handles to it, to permit of its being raised up with each course, and moulded around again.

At every new course in the construction of the walls, slips of rough boarding, an inch thick and six or seven inches wide, should be laid flat on the bottom of the mould, to be worked around and completely bedded in. These slips should be laid alternately, so as to break joint, and made to act as binders to hold together and stiffen the walls. At the angles they should be fastened together with oak pins, projecting on both sides beyond their combined thickness.

These binders should be completely covered, so as not to let the air or damp approach them. Nails should not be used in fastening them, as they rapidly rust and cease to hold.

The gables may be constructed in the same manner as the walls, and be raked down evenly, after the crowning-block is on and the mould removed.

The chimney-shaft may be constructed of alternate layers of earth, and wooden slats bound at the angles by long wooden pins penetrating through the entire height of the chimney-shaft above the roof. The whole to be carefully plastered within and without. The lower ends of the four binding-pins might be secured to the roof at the rafters; or, if carried lower, might be fastened to the collars of the roof, which would enable them to resist the storms.

The capping-boards should be amply sufficient to protect the outside from the falling rain. For which purpose, also, it should have a groove around it on the underside, to prevent the coursing of the rain-drops along its under-surface, and down the outside of the chimney-shaft.

The Roof.—The walls and gables being now up to their top-level, square pieces of wood, nearly as long as the wall is thick, may be laid across at every two feet apart, all around the walls and gables, and be bedded in the clay. On these may be laid and secured the wall-plate, which is a connected length of two-inch boards, on which the roof is to rest. That is, the rafters are to be notched on their underside at about two feet from their ends, and at this notch they are to be pinned or secured to the wall-plate. A stick, say two inches by four, may now be laid across the tops or points of the gables, and stretching over the whole length of the house, project two feet or more at each end. If one stick cannot be had of the required length, let two, or even three, sticks be halved together to make up the required length. Against this pole-plate, or ridge, the rafters are to abut and be nailed. It will, of course, be divided where the chimney-shaft comes through the roof. At this point each piece will be nailed to the rafters next to the chimney. But perhaps the simplest construction would be to omit this pole-plate altogether, and nail the rafters across each other at their tops. Sawed rafters are, of course, the most desirable; but when they cannot be had and only boughs of trees are available, they must be as straight as possible; the side which is to lie uppermost must be chopped as even as possible. Take a piece of board, and cut it into the required form of the rafter, making the notch where it is to come, and cutting the lower end square off. This is the mould by which all the rafters will be uniformly cut. In order to make this mould correctly, it will be necessary to lay out, or mark, the plan of the roof, full size, on the ground in this manner: Measure half the width of the house to the outside of the wall, and add the projection of the rafters beyond the wall, say a foot and a half, to that. Let us suppose the whole width, from out to out of the walls, to be eighteen feet; adding twice one foot six (the projection), it makes twenty-one feet; the half of which is ten feet six, the height or pitch of the roof. Lay off this pitch on the ground, and with a rod, or straight strip, join the two points (the top of the pitch and the extremity of the breadth), and the length of the rafter is found; that is, fourteen feet nine inches, nearly. Now mark where the notch will come, and square it off. Apply the piece of board you intend for your mould to this, and mark it accurately; then cut it out, and from this mould mark alike, and cut all the rafters, taking care that there are none of them less in length than the required standard they may overrun, for it is easy to cut them off afterwards when nailed together in their places.

Where the chimney-shaft comes up, nail on bridging-pieces, from rafter to rafter, at each side of it, and as close as to help brace it.

Board over with rough boards the whole roof, and shingle it. Stop up all the sleeper-holes by ramming clay into them on both sides.

Now, plaster and rough-cast the outside, and plaster the inside of the house; of course, giving sufficient time to the fresh filling up of the holes to dry.

When the walls of a compressed clay-house are erected in the summer, they should not be plastered until the fall, so that the action of air in that time may take the moisture completely out of them. And if built in the fall, they should not be plastered until the following spring, or until the frost is quite gone.

To prepare the walls for plastering, a small scaffold can be easily constructed with a few poles and some boards; taking care not to bear upon the wall. The walls should be first indented or picked with the sharp point of a hammer, all the dents being made as close as possible to each other, and each being deeper above than below, so that the plaster may have a good hold on each dent. Previous to putting on the plaster, let a stiff brush be applied to the dents to remove any loose earth or dust.

The *rough-cast*, or pebble-dashing, which is to go on the walls, is composed of lime and sand, in equal parts, sifted through a coarse sieve and mixed with water in a tub, to the consistency of cream. The wall has first to be sprinkled with water, either by a brush, or a bundle of twigs, to prevent the surface receiving too much wet.

The rough-casting must be carefully done, by first sifting the sand away from the stones, holding the sieve in water, then running the grit through a coarser sieve, likewise in water. To this grit as much lime as will make it hang well together is added, mixed with clean water well stirred in a tub; and in putting it on, a coat of hair-mortar is spread over the wall, and the rough cast taken up, a little at a time, in a small shallow tin pan or shovel, and thrown against it.

If hair cannot be readily had, some hay or straw, or split reeds, cut in short lengths, may be used.

In building the chimney, it might be easier to have it in the corner, at the junction of the partition with the main walls; but it would be a very undesirable position, when we remember that a large percentage of heat would be lost on the outside. Besides, the flue would not draw so well, having its back exposed to the weather; and the chimney-shaft coming through the roof, near the eaves, would require a special gutter between its upper side and the slope of the roof, to catch and convey away the rain-water, which matter, if not very carefully attended to, would be sure to cause leakage and all its disagreeable consequences.

THE USE OF BRICKS WITH EARTH.

Where bricks can be procured, even in small quantity, it would be very desirable to use them in building the chimney, and especially the shaft above the roof. It would likewise be found advantageous to turn brick arches over doors and windows. The clay, or rammed earth, adheres to them readily. For the jambs or sides of windows and doors, bricks are very desirable, as combining so well with the rammed earth.

When bricks are used with the rammed earth, they should be laid a short time before the earth is rammed against them, to allow of the proper setting of the mortar, which should be very good, and not be mixed with much water.

In concluding the subject of compressed-earth building, we must urge its superior claims on the pioneer, whose object is to provide for his family a dwelling, at once permanently strong, comfortable, and healthful; while it is economical in construction. It may be objected that it takes a great deal of time and labor; but if we look closely into the subject, we will find that the time required is not at all equal to that wasted in the drying of more moist materials; and that the labor really does not exceed that expended on very inferior modes of building. Walls of compressed earth are thoroughly dry in forty days, and may be relied on to bear the weight of the roof at once, on completion of the last course.

It must be understood that these walls dry as well in winter as in summer; for it is not the action of the sun's heat so much as the influence of the air that produces the desired effect.

ADOBÉ, OR SUN-DRIED BRICK.

HAVING described the process of making walls of rammed or compressed earth, and recommended it as the best mode of using earth as a building-material, we will next proceed to explain the nature and construction of what is called *adobé*, or sun-dried brickwork.

An area of ground which possesses the best description of earth for the purpose is laid out, and the top surface removed. It is then dug up, or ploughed, and water turned in on it; previously scattering over it thickly, either cut straw, reeds, sedge grass, or any convenient *binder*. This whole area of broken ground is to be well trodden by oxen, until the binders are thoroughly worked in with the wet earth, and the whole has become a thick *mud*, through which the straw or sedge has been well mixed.

Now have ready, on a wide board placed near this *mud bed*, say six boxes made of inch-boards; each box being eighteen inches long, twelve inches wide, and six inches high, on the inside. Let the two sides of each project two or three inches beyond the ends, and to the ends nail cleats, rounded on the outside, to allow of their being easily handled.

Fill these boxes, or moulds, with the material, and even off the tops with a straight-edged stick called a scraper. Then commence with the first one filled, and taking hold of the two cleats, raise it off. This do with the others; and when all the blocks are uncovered, trim them, and remove the plank, bearing them to a convenient spot where the rays of the sun can strike them; and place a fresh plank, with the boxes or moulds to be refilled. Previous to using the moulds, let them be washed on the inside, to prevent any dirt adhering to them and injuring the new batch. If the sun is very powerful, it would be advisable to form a little pond of water in which to soak the moulds, and keep them from splitting.

The whole number of blocks required for the building of the house should be cast, and left to dry before the walls are commenced; and if the number is large, the first-made can be used soon after the last block is cast. But care must be taken that they are pretty well hardened before they are set in the wall. While drying, they should be turned upside-down; and when about to be used, they should be carried to the wall on the same board which they dried upon, and be slided carefully into their places. This will avoid breakage, and greatly facilitate the work of construction.

In the use of these sun-dried bricks, due patience is required, so as not to enclose the window- and door-frames before the bricks are perfectly dry, as these walls shrink seriously after they are up, and would in consequence cause much damage to the structure, if the frames were permanently fixed before time was given to settle. It will be necessary, therefore, to insert temporary rough frames until the walls are thoroughly dry, to prevent trouble. These will then be withdrawn and the regular frames inserted and secured.

Walls of sun-dried brick should be build like coursed masonry or ordinary brickwork, and the joints and beds should be of lime-mortar, or, if lime is not to be had, of clay and sand mixed with water to a stiff consistency.

The foundation-walls should be of stone or burned brick, carried up at least twelve inches above-ground; and in case there is an absence of lime, they may be made of broken stone packed with gravel and sand, the whole levelled off on top, and, if possible, covered with flags, flat stones, or slate, to cut off the rising damp from the walls.

The door- and window-frames are to be made of stout plank, as broad as the wall is thick; and thick slips are to be nailed up, back and front, to cover the meeting-joints of the brick and the wood, as, in drying, the shrinkage will cause a wide opening. It would be advisable to fill up and ram this part of the building very well, using as much gravel as possible.

These frames should be attached to oak sills, the under outside edge of which should be grooved, to cut off the coursing of the rain against the wall.

Hardwood pieces, two or three inches thick and of the width of the wall, should be laid over all the window- and door-frames as lintels. To these at any time may be nailed hoods or weather-strips for protection and ornament. In all cases, they should project boldly beyond the walls, and have their outer and under edge grooved in a similar manner to the oak sills.

The wall-plates, for the roof to rest upon, should be stout, and be securely halved and pinned together at angles.

The chimney ought to be constructed of burned brick or stone, and if neither of these materials are to be had, the directions already given for earth-chimneys may be adopted.

The walls of these sun-dried brick houses are plastered and gravel-dashed, or rough-coated, on the outside, and skim-coated on the inside, in a similar manner to the rammed earth walls before spoken of.

Although we do not consider these *adobé*, or sun-dried brick houses, at all equal, either in strength or comfortableness, to the rammed-earth construction, yet they answer their purpose very well, are warm in winter, and cool in summer. They are to be found in all parts of South America, and many towns are entirely built of them. Even in northern latitudes they form a good shelter, and have been tried many years ago in Canada; not being equal to burned brick, however, where that material was available, they could not become popular there. Yet there are conditions under which the sun-dried brick is a desirable mode of construction as being ready and cheap; and if not used for building the dwelling, may be very judiciously applied to the erection of the barn, or the yard walls, requiring for the latter use to be well covered on top, and securely sheltered from the weather.

During the raising of the walls of sun-dried brick, as well as those of rammed earth, it is very necessary, as a protection against rain, to have shelter-boards ready to lay along the top of the new work; and this precaution should be taken every night until the roof is on.

GRAVEL OR CONCRETE CONSTRUCTION.

WE now come to the subject of building with gravel and lime, where these materials are handy, yet stone is scarce, and where wood is difficult to be had. In certain localities, such for instance as our prairies, limestone beds, as well as underlying strata of gravel, are to be found; and with such provisions of Nature, man can surely furnish himself with a dwelling equal to any in comfort, durability, and economy.

It has become usual in cities to employ water-lime, or hydraulic cement, in the making of concrete, and it is no doubt the most desirable for the purpose; but common quicklime can, nevertheless, be used very efficiently in the casting of walls of great strength and durability.

The great object is, to hit off the true proportion of the gravel and lime, to ensure their perfect combination, so as to form a truly concrete wall. The common mistake is, to make the composition too rich by the introduction of too much lime, and this error produces a weak wall, and one very apt to swell and crack. The opposite error of putting in an undue quantity of gravel is not so bad in its effects, and if sufficient ramming be applied, it may still make a very strong wall.

The foundation for the concrete house should be formed of the largest stones, laid flat in the trench; and over them stone chips, filled in with gravel, and the whole covered with a creamy mixture of lime and sand. Build the foundation high enough to keep the walls above the possibility of damp from the ground. The foundation-walls being fairly levelled up, the framework for the casting of the superstructure is to be set up in the following manner:

Take sixteen stout posts, say twelve feet long, and set them up, four at each corner, forming a square. Each of these squares is to include the thickness of the walls; sink them in the ground about three feet, wedging them well down, and bracing them well together above, at their tops. Now insert couples of such posts along the length and breadth of the walls, between these corners, making their distance apart from each other, say four feet. Brace and secure these well; and insert on the inside, of the exact thickness of the intended walls, two sets of boards on edge, not less than one and a half inch thick each, and say twelve inches broad. Keep them apart by means of sticks as long as the wall is thick. They must be smooth on the inside, and free from splits, as they mould the wall, and on their surface depends its good or bad appearance.

Take care to set up, and secure the door-frames at once. Then proceed to cast in the concrete, which should be prepared as follows:

In a large mortar-box, capable of holding twenty bushels of concrete, place fifteen bushels of coarse sand, or gravel, with small chips. Now mix (in a small box) about one bushel and a half of lime; slaking and reducing it with water to a thick cream. Into this pour fine sand, sufficient to let the lime-water or cream of lime be thickly fluid; and pour the well-mixed contents of this box well over the large one. Lastly, work the mass thoroughly together, and as rapidly as possible; adding as much fine sand as it will bear.

Now cast this concrete into the space between the boards, commencing at the door, and going around the walls, spreading it only four to six inches thick, until the first course is all laid, and the boxful exhausted. Proceed as before, and so on, layer over layer, until the work of the walls is completed. As the enclosing boards forming the wall become filled, lift them up higher, and again secure them as before. There are two ways of proceeding in this matter: One is to loosen and raise the boards higher, supporting them on upright posts cut to the proper length, leaving a good lap on the finished work, in order to prevent the occurrence of any appearance of break. The other is to let the boards just used retain their position, and place fresh boards above them. The advantage of this latter plan is, that, by not uncovering the freshly-made wall, it gives it a chance to harden while the next course is being laid. The chief objection to it is, that it takes double the quantity of boards. However, as these boards will all be available in the dwelling, it really makes no difference on this point.

The builder cannot exercise too much caution in the matter of *plumbing* and *levelling*, as it is impossible to correct any defects once the walls are up.

The openings for the windows and doors should be cased with plank, an inch and a half thick, and of the width of the wall; within which the regular frames for the windows and doors are to be accurately set, when the walls are all up, and not liable to changes incident to drying.

The chimney-stacks should be made the usual size, that is, two feet thick; and be carefully incorporated with the walls. The flues might, with very great advantage, be made circular on plan, by using a round block, or a tin cylinder with a handle to the top of it, to enable the workman to draw it up as the flue rises in height. Some persons prefer inserting stove-pipes, and moulding them permanently in. But this plan is objectionable, as these stove-pipes will inevitably wear out, and cause obstructions in the flues, often very difficult to remedy. Besides, where it becomes necessary to introduce a bend in the flue, the movable cylinder would be the only easy method of accomplishing the purpose.

As regards the thickness of the walls, we would recommend their being twelve inches throughout, when the house is but two stories high; but when three stories, the first and second might be sixteen inches, and the third twelve inches in thickness.

Over all openings for doors and windows, there should be a lintel of seasoned oak or ash, not less than four inches thick; or two planks, of two inches in thickness each, laid flat. This will serve to bind the building at the openings, which are the weakest parts of the walling; and likewise to guard against the bad effects of any jarring or tremor of the floors above them.

If there are gables, they should be constructed in the same manner as the walls; the sloping courses being the only difference, and these are easily managed.

BÉTON, OR CEMENT BLOCK CONSTRUCTION.

THIS mode of making concrete is of French origin, as the concrete itself is of English invention. The difference lies chiefly in the use of water-lime cement, instead of common lime, and the rapidity with which the *béton* sets demanding a different manner of moulding or casting it.

Water-lime, when moistened and formed into a ball, may be thrown into water, and left there, it growing at once so hard that water has no effect on it; in fact, it resists water. Common lime, if treated in the same way, would immediately dissolve and mingle intimately with the water. Common lime, therefore, requires the aid of sharp sand, or gravel, to give it consistency and strength; while the addition of sand to water-lime weakens it, just in proportion to the quantity of sand used.

We think it right to thus put the constructor in possession of the plain facts, so that he may go to work intelligently, and use his own judgment and inventive skill. Comparatively very little is known on this subject; and it is but a few years since our best architects and builders used to shrug their shoulders and smile incredulously at the very name of *concrete:* nowadays, the foundation, which has not concrete in its composition, is an exception to the general rule in building.

In presenting this subject of *béton* block construction

in a work like this, which is intended exclusively for a class of persons whose means are very limited, it must be presumed that water-lime is to be had cheap and plenty; for there are many places throughout this country, where it has been liberally supplied by nature, and such localities will not be long without competent and experienced hands to prepare it for its purpose. We must also assume that a brick-yard is near enough to afford a supply of its refuse or waste material. These, with the use of sand and gravel, give all that is necessary to mould and make the best walls that a house can have.

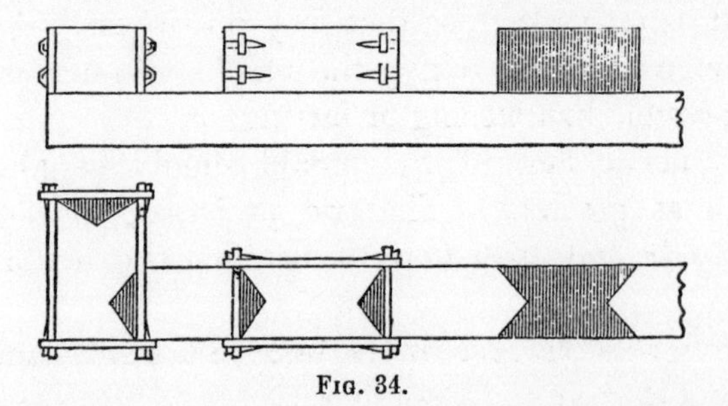

Fig. 34.

A reference to the accompanying illustration will at once show the form of each block as it is cast, and the mode of spacing the blocks.

The manner of making the mould itself is simply as follows: Take two-inch and a quarter boards, say twenty-four inches long each, by ten inches high, and about three inches from each end, cut a square groove one inch wide, parallel with the end, not less than half an inch deep. This groove is made thus: Having marked the place for it, take the saw and give two cuts square across the board, and half an inch deep. Now chip out the wood between these two cuts with an inch-chisel, and make the groove even and clean. Do the same at the other end of the board. Groove the second board in like manner. In each of these grooves, near the top and bottom, cut a hole through, say two and a half inches long and an inch wide; and the sides of the mould are made. Next prepare the two ends, each to be twelve inches long, ten inches high, and an inch thick, taking care to have two projecting pieces to fit into the corresponding holes in each groove, and each of these projections, or *lugs*, as we will call them, have square holes in them, into which square pins can be fitted close up to the sides, thus fastening the sides and ends of the mould together. These pins should be tapered off, so that they can be easily loosened and removed, when taking the mould asunder, with the least possible hammering or jarring.

The inside face of the mould should be planed as smooth as possible. The two projecting sides should likewise be cut with *lugs*, to facilitate the handling of them.

The figures above show the form of the sides and ends, and how they are put together.

The great advantage in moulding in blocks is, that it does away altogether with the necessity for upright posts, and all the unwieldy arrangements necessary for the casting of concrete by the old method, already illustrated. Neater work can be done, and any error or irregularity at once detected and remedied. It may be said that these moulds are expensive, and that when the house is completed they are useless. But such is not the case, for these moulds will prove useful where there is always, as on a farm, a want arising of some addition or outbuilding; and it is easy to keep them in order, by setting them on a temporary bottom, and keeping them filled with water. They can pay for themselves, too, by being hired out to neighbors desirous of building in the same manner.

The moulds, being all ready, are placed around the site of the walls, as shown in the accompanying figures, and every mould in succession is filled in the following manner.

The *béton* is put into each to a depth of three inches, until the last mould is supplied. Each mould-layer is to be then well rammed, or pounded, until the *béton* is perfectly hard, when another layer is to be supplied to each mould, and is likewise well rammed; and so on, until the moulds are all filled up to the top. They are then allowed to stand an hour or so, to let them shrink a little in drying, when they can be loosened by gently driving out the pins, or wedges; then take off the sides and ends; put them all together again, and place them convenient to the walls. Meantime, the spaces between the blocks will be enclosed at each side by smooth boards, held securely in their places. These boards should be long enough to lap well on the new-made blocks. The mode of securing these sides to the blocks, while filling in and packing the spaces, is as follows: Take four straight pieces, three by four inches, (boughs trimmed and axed square,) and let them be, say three feet long each. Take two pieces of oak or ash, five inches wide, two inches thick, and thirty inches long each, and cut two holes in them three by six inches each.

The four three-feet pieces, already spoken of, being set upright at each side of the adjoining blocks, these two collar-pieces, just mentioned, will be put on over the tops and brought down until they bear upon the blocks. The sides will now be let in between these, and resting against the blocks. It will be seen that the holes in the collar-pieces are each six inches long, and that the uprights are but four inches; thus leaving two inches of space to be filled, in order to lighten the collar on the uprights. This is done by two wedges for each hole, one set in above, and one below. It only requires two stays or sticks at the head of the uprights, of the necessary length to press them apart and cause them to tighten below, and thus firmly hold the side-boards in their places. Two small blocks may be tacked on the inside of the uprights to prevent the cross-pieces from slipping down.

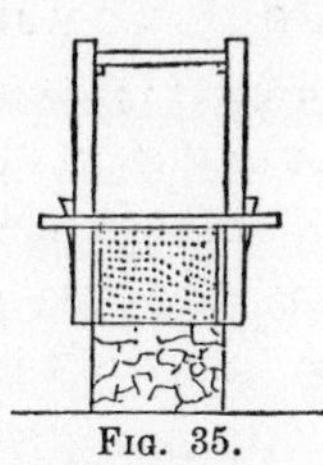

FIG. 35.

The Beton Mixture is thus prepared: — Put together a mortar box six feet square and ten inches or a foot high. This will be large enough to make eighteen feet of wall in continuous *béton*, and will give material enough for a day's work for two men. It is necessary that no more stuff should be mixed than is required for the day, as any remainder will set and be useless.

Take of well-sifted, clean sharp sand, four bushels; add to it one bushel of cement. Now dissolve lime in the water you are about to use, so as to make a thin cream, and proceed to mix up the cement and sand with it. Now put in such broken stones and broken bricks as you can collect, and work the mass together. Or, put the mixed cement and sand into the moulds and afterwards pack in the broken stone and bricks, keeping these to the centre or heart of the blocks. If the sand be very fine, the proportion of cement must be increased; as, say three or three and a half of sand, to one of cement. It must be thoroughly and rapidly mixed and thrown into the moulds, and each layer, as has been already said, must be very well pounded, until the whole forms one compact mass, or concrete block.

The object of inserting the triangular pieces, seen in the engraving, is to produce dovetail joints, and thus bind all the work together in a perfect manner. Rounded corners can be made without any more cost than the trouble of inserting a hollowed block in the proper angle of each corner mould of the outer wall of the house. The effect is good, and worth the trial; the amount of material saved at these four angles will more than pay for the trouble of inserting the blocks.

Where windows and doors occur, it is better to place the moulds next to them, and let the jambs take the place of the ends; previously tacking on an inch and half slip to the back of each jamb, so that the *béton* blocks may have a good grip of the jambs when the walling is completely cast. This will also prove a weather-tight joint. In covering the lintels of doors and windows, the material must be very well worked so as to become like putty, and three layers, of three inches thick each, must be carefully spread on and beaten until so hard as to give forth a sound under each stroke. No stones to be used; only sharp, perfectly clean sand, and pure cement, with only as much lime-water as will cause it to mix and form a stiff composition.

If a pug-mill, such as is used by brickmakers and pottery men, could be put up for the purpose of working or kneading the cement and sand in the soft state, it would insure a *béton block* that would rival the hardest and best sand-stone.

THE OVEN.

ONE of the most important aids to home comfort is the oven for baking bread. It is always built outside of the house, and of such materials as the case affords; but brick is of course the best for the purpose. Supposing there is no brick to be had, it then becomes necessary to construct the oven of earth in the following manner.

First erect a rude platform or table, two feet high, of logs covered with slabs, so as to form a square of, say, four feet each way. On this platform lay out a floor two feet square, composed of clay one part, earth two parts, and fine gravel three parts, previously mixed up with water, and well worked into a very stiff paste. Lay it on three or four inches thick. Let it dry, after first beating it with flat sticks. Then, when dry, pile up on it, with a basket or box, a mound of sand, wetted to keep it in place. This pile of sand should be two feet high, and rounded on the top, much like a bee-hive. Now mix up similar material to that of which the floor is formed, but enough of it to completely enclose the mound of sand with a shell at least four inches thick. A short block of wood, round or square, and about one foot each way, should be laid, for the doorway to be moulded around, on a level with the floor. When the shell of the oven is completely cast, then sod or cover it with green turfs, and leave it to dry. After three or four days withdraw the block at the doorway, and let the sand escape. Clean the interior out, and with smooth boards go over and make the interior even. Spread over the floor a thin layer of fine sand, and beat it down evenly. Lastly, put into the oven a few handsful of twigs, and set fire to them. The door may be a thick board with a cleat across it, to give it strength and serve for a handle. This door should fit well against the oven, and be secured, when the bread is baking, by a large stone placed against it; or, any other convenient means. Of course, the thicker the wall of such an oven is made the better. The final covering of sods of grass should be made so sloping as to give a ready current to the rain, and if this oven were built under the shade of trees it would be all the better. The object in raising it on a platform is to keep it dry, and at the same time to save those who use it the labor of stooping.

If each layer of earth and gravel is well mixed and beaten hard, there is little doubt but it will prove a strong oven, and serve its purpose until such time as a brick one can take its place.

The Dutch settlers build such ovens, and continue to use them for many years.

THE CELLAR.

THIS is one of the most desirable appendages of a house, but is too frequently located in the basement, giving rise to damp and foul air most injurious to health. No matter how well drained a cellar may be, the fact of its being under the house and below-ground is sufficient to insure its unhealthfulness. In cities there is mostly a positive necessity for this subterranean location, but in the country there is no such necessity, and therefore the cellar should be always removed from the house, yet still conveniently near to it. We will now describe a cheap cellar.

In forming the cellar, the entrance should be located at the south end of the proposed site, and as much defence or security against the severity of the northerly

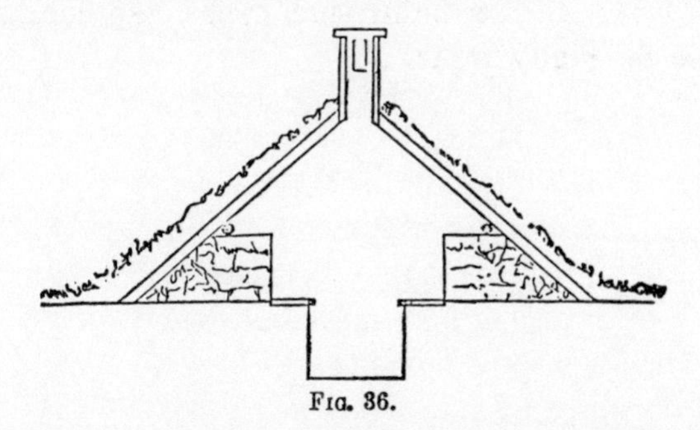

FIG. 36.

blasts had as possible. Supposing the cellar to be six feet broad and eight feet long, dig out three feet and a half by six feet, and three feet deep. Cut a quick slope three feet wide for the descending steps to the entrance. Now commence to build a dry wall of such stones as can be had, filling in between them with gravel, and let this wall be on the outside of the six feet by eight, already mentioned. Make it two and a half feet thick at the base, straight on the inside, and sloped or battered on the outside to a height of two feet; being one foot thick on the top. All around place heavy split trunks of trees, halved into each other with the axe, at the corners, and lying, face downwards, on the walls. Now, crossing these and leaning against the outside slope, lay slabs of equal length, say ten feet long each, meeting at the top, and resting on a straight trunk laid from gable to gable, and held up at each end by stout boughs set crossing each other at top, and well tied together. The flat side of the slabs is to face inwards. Cover these slabs with others, breaking joint throughout, that is, covering the joinings of the first set of slabs. Next, shovel on gravel and clay, say six inches thick, and then sod it all over with good rich sods at least four inches thick.

At about the middle of the roof a square box, made of rough boards, should be set up and fixed as a ventilator. It should be eight or ten inches square, and about four feet long, covered over at top by a cap projecting three inches beyond its face, all around. In each side of this ventilator cut a hole, twelve inches long and five inches wide, and tack over these holes gauze wire, or open canvas such as is used for very coarse strainers. It would be well to have the cap of the ventilator movable, so that in hot weather all the ventilation possible may be obtained. On the inside, it will be observed, ample provision has been left for shelving; for the difference between the width between walls (six feet) and the width of the dug-out part (three feet and a half) leaves eighteen inches wide on each side; and the length being eight feet from front to rear, and the dug-out part being but six feet, gives two feet deep for shelf at the opposite end to the entrance. Besides, the top of the walls being two feet thick, and having only to give place to a trunk, say ten or even twelve inches diameter, will leave twelve inches bearing for the upper shelves. These should be overlaid with boards planed on the upper side, and nailed down to short thick sleepers bedded crosswise in these spaces.

If it be deemed necessary to strengthen the bearing of the slabs of the roof, purlins, of thick boughs, can be laid half-way up the sides bearing on each end, and sustaining the slabs midway from the top of the wall to the top of the roof.

The finish of this cellar, if boards were to be easily had, would be to sheet up the sides and end; but if boards were scarce, the sides and end might be faced up with plaster of earth, sand, and wood-ashes, mixed with water, in which, if possible, lime is mixed to the consistency of a thick cream. The floor may be neatly paved with small round stones about the size of a turkey-egg, such as are found in gravel-pits, or by the side of streams.

Particular care must be taken to form a good drain around the outside of the cellar, which, if the ground has no fall to it, should empty into a pit, or a sunk barrel at a safe distance from the cellar.

In winter, if the frost is very severe, it would be

judicious to cover the cellar-roof with a good thick layer of straw or rushes.

The Potato-Pit.—Before the frost sets in, the potatoes should be dug, and they should be left a few hours to dry. This time may be occupied in digging a pit large enough to contain them, in a dry spot (sandy, if convenient). A layer of dry straw is put in the bottom of this pit, and a layer of potatoes over it. Another layer of straw, and another layer of potatoes; and so on up, until the potatoes are all pitted. The earth, which was taken out of the hole, is then to be banked up and over the pitfall, and the whole smoothed down in the form of a roof with a sharp pitch. A drain, or gully, is to be made (as in the case of the cellar), and all is complete. This potato-pit will have to be opened occasionally, in fair weather, to have the shoots rubbed off with the hand. Great care must be then taken, in closing up again, not to leave any cracks by which wet may enter, or frost take effect.

FENCE-POSTS AND GATE.

IT is always advisable to build a house back from the road at least sixty feet, so as to admit of a front plat or flower-garden. Some build on, or close to the road-side; but this is not at all desirable, as pigs, poultry, &c., will make it a hard matter to keep the front of the house anything like clean; and the fragrant beauty of the flower-garden is worth all the pains and trouble that may be bestowed on it. A cottage home with a neat exterior is like a well-dressed person; it draws attention to itself, and the possessor is justly proud of it.

Fences, or walls, must be made of whatever material is readiest to the pioneer settler. If stone is plenty, a dry stone wall is the best. Lay the largest at the base, and break joint as much as possible; filling up all holes with small stones and gravel, so as to leave no harbor for vermin. Let your wall be wide on the base, and sloped up on the inside, keeping it perfectly perpendicular on the outer face. Earth and sod the slope evenly. This will make a permanent wall, and always look well. If more defence is needed, it is easy to plant the ridge, or top, with prickly shrubs, thorn, &c. Four feet should be the height of the stone-work, and it should have a sunk foundation of twelve inches well rammed.

But if stone be not plenty, or rather scarce, and wood is abundant, then rail-fencing is perhaps the easiest of accomplishment. What are called snake-fences are a slovenly makeshift, having nothing to recommend them save the facility with which they are constructed.

The following mode for setting posts for fences is perhaps the very best which can be adopted:

Dig holes two feet square and three feet deep, six or seven feet apart; cut a three-inch slab into squares of eighteen or twenty inches square each; and in the centre of each of these squares cut a hole six inches square. Now, with the axe, square down the end of the post to fit tightly in the hole. Set the square board, or slab, in the hole, and fix the end of the post into it. Then drive the post down firmly into the ground below the square board; and, lastly, fill up, and ram around the post. Each post is to be so sunk in a similar hole, and anchored to a square board in like manner. It must be evident that nothing can shake the posts so set or anchored; and they will bear any amount of stress on them. These posts may be nine or ten feet long, leaving six feet over ground when they are fixed in their places. Now for the placing and securing the rails: Supposing the post, from the anchor-board up, to be eight inches wide, begin about four inches above ground, and, with the saw, make a cut sloping inwards about three inches. Now chalk or mark off a straight line from the end of this cut to the top of the post, and beginning at the top, saw down along this line to the end of the sloping cut first made. Remove the piece, and on the inner face of the post just laid bare mark off, at four inches from the bottom of the sloping-cut, a line across the thickness of the post. Four inches above this make another mark, cut a kerf an inch and a half deep, with the saw, at each of these marks, and chisel out the wood. Six inches higher repeat this operation; and eight inches above that, and twelve inches above that again, and once more at fourteen inches above this last make similar cuts. Treat all the posts in like manner, and fit the rails to them. If the rail should be thicker than the cut is deep, then cut the rail itself to match it. Next replace the piece sawed out, so as to cover the rails. This piece will fit tightly into the kerf below, and can be secured above by a square cap, to fit down on it and the post, binding them together. No nails are required in this mode of fencing, and it has the advantage of being easily taken asunder, and put together again to replace a bad rail, or the like.

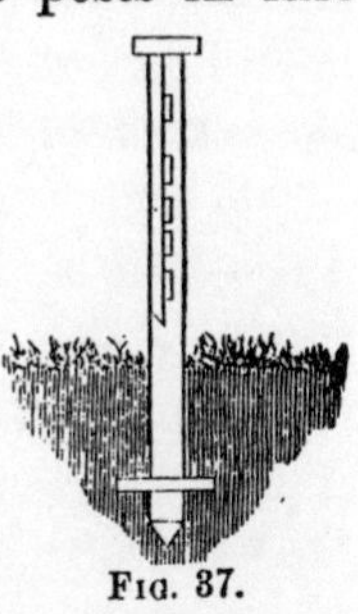

FIG. 37.

The Gate-Posts can be made and set up in the same manner.

The gate itself should be made of inch-boards nailed to a stout frame, and the whole braced together with boards crossing each other (one on each side), and nailed at the angles of the frame. The side of the frame which is to be hung is called the *hanging stile.* This should be longer by six inches, both above and below, than the other stile, or side of the gate, and these projections should be made round and smooth; the lower one might be cut to a point, and bear upon a piece projecting from the gate-post, having a shallow hole for it to work in, so as to relieve the post of part of the weight. The rails may be let in to the two sides, or stiles, by the same mode as that used for the fence,

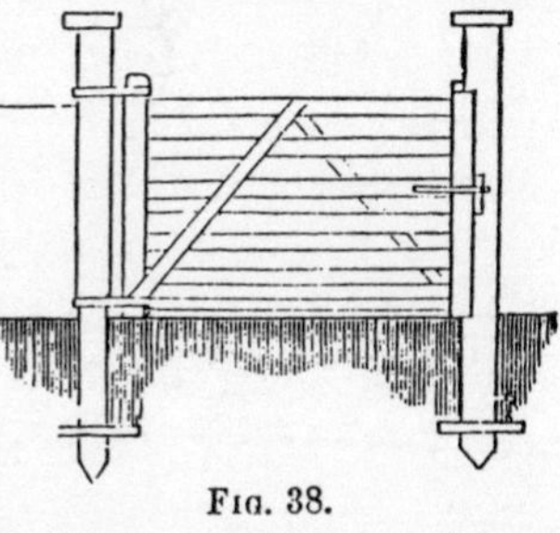

FIG. 38.

already described, and a cap, or binding-piece, be fitted down tight on it, as far as the top rail. From that up may be rounded; the two parts will be held sufficiently close together. A similar binder should be fitted on below, up close to the under part of the first rail.

The hanging of the gate may be done in this way:—Take two boards of oak, or any tough wood, two inches thick, each fourteen inches long and eight inches wide; cut a hole four inches square through at one end, and a round hole four inches in diameter at the other end. Fit the square hole in each of these on the gate-post, above and below, with the round holes to receive the hanging stile of the gate, and work loosely, so as to open and shut with ease, (Fig. 39.)

Fig. 39.

ECONOMIC FURNITURE.

NEXT to the building of the house is the furnishing of it, and as necessity is the motive throughout, so it must govern in this matter, calling forth all the inventive genius of the handy-man to provide comforts for his family out of the humblest of means; and this can be accomplished sufficiently well to suit the purpose with the exertion of the merest instinct.

THE BEDSTEAD.

The most desirable article of furniture is that on which to rest the limbs after a hard day's work; and here it may be well to say that the custom of laying a bed upon the floor is a very unhealthy one, as well as being slovenly and dirty. The foul air accumulating in a room tightly closed for the night, being heavy, will settle down to the floor, and it is evident that the lungs placed at that level must imbibe all its impurity by inhaling it in the act of breathing. The soldier, who enjoys sleep in the open field, is free from the injurious action of the confined gases which the sleeper in the close room has to swallow.

The bed should be always raised not less than eighteen or twenty inches above the floor-level, and the frame or bedstead which effects this desirable purpose should not be connected with the wall or floor, but be movable for cleanliness' sake alone. However, where a sleeping-place is very narrow, advantage may be taken of the fact, by nailing pieces to the walls, and laying the slats loosely on them. In such case, the slats being easily removable, the process of cleaning will not be interfered with. But in the larger room the independent bedstead must be constructed, and that is effected in this manner:

Take two pieces of hard wood, each four feet long and four inches square, and in each insert three legs of hard round wood. Let them be sixteen inches long, and four inches diameter; square them off evenly, with the saw, at the ends, so as to make all exactly equal. Now cut a square four inches deep at one end of each, and cut a hole in the four feet-pieces to match these square ends, and drive them tightly together. To the middle attach an offset, as a support, as seen in Fig. 40; but none to the outside legs. This will enable the three-legged pieces to stand alone. One is the head, and the other the foot of the bedstead. Place them, say six feet apart. Next take two-inch boards, fourteen or sixteen inches wide by five feet long, and plane them up neatly, making the edges even. About the middle of the width of each of these, cut two holes, three inches long and two inches wide; and, lastly, prepare two side-pieces, seven feet long, three inches thick, and four inches deep; cut six inches of the ends of these to fit the holes in the two boards (the head-board and the foot-board), bore a hole, and insert an oak pin in the two long pieces, or sides; close up to the outsides of the head and foot-boards, and the bedstead is framed together. Now cut, in the top-edge of each side-piece, square recesses or beds for the cross-laths to lie in. These may be seven or eight

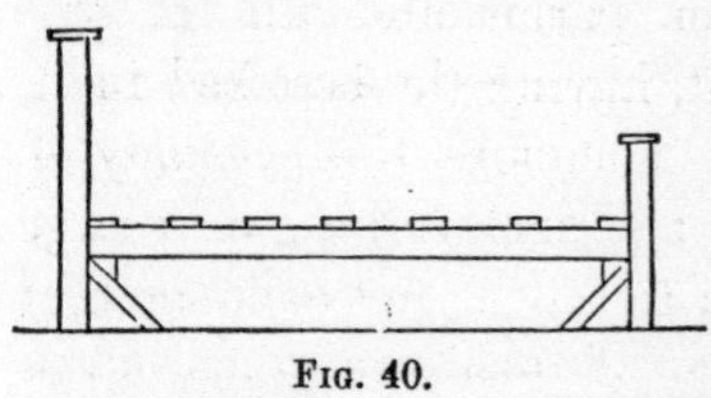

Fig. 40.

inches from centre to centre and about three inches wide. The laths should be of ash, three inches wide, four feet long, and three-quarters of an inch thick. They should fit nicely in their berths.

This makes a substantial bedstead, and one easily taken asunder, by removing the oak pins at the head- and foot-boards. Any handy-man can make it without using a nail or a screw, and a little taste displayed on the legs and foot-board may make it quite presentable for a rough cottage-home.

THE TABLE.

In small houses, such as these we describe, the living-room is kitchen, dining-room, parlor, and all. It generally is a bed-room by night, having the best bed in it, for the father and mother. Therefore, the economy of space is a very great object; and such a thing as a large table, occupying the centre of the room, not to be thought of for a moment. This article of furniture, when not in use for meals, is set aside against the wall; and even in that position it takes up too much space. By observing the following simple directions, the handy-man can easily construct a table quite suitable for the wants of his family, which will not be in the way when idle: Let him saw off a perfectly straight limb of any tree. This should be five feet six inches long, and about three inches thick. Saw it in two equal parts. Saw each half down the middle of its thickness, and clean up the faces smooth. Now saw off another limb of similar length and thickness, and treat it in exactly the same manner. The four sticks thus prepared are to form the legs of the table, and they are to be put

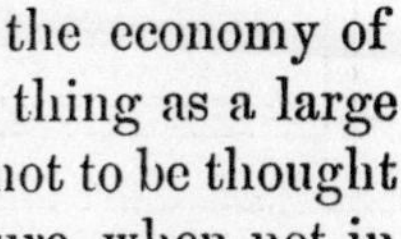

FIG. 41.

together in the form of two X's, (see Fig. 41): Place the first pair upright, and two feet apart. Now draw the upper ends across each other, until those upper ends are eighteen inches apart; and where they lap one another, mark with a pencil or a nail. Take them apart, and lay them on the ground. Now in the centre of the square formed by the pencil-marks, bore a hole in each, taking care to have them exactly opposite each other, and do precisely the same with the other pair of legs. The holes thus bored should be at least an inch in diameter, and two round pins should be made, to fit them fairly. Then work the legs, with the pins in, backward and forward, until the pins are sufficiently loose to admit of the legs being folded together with ease. Put each pair of legs together, with the pins in, and stand them up. Now, with a straight edge, square them across, top and bottom, and saw them off smooth. Next, cleat together three boards, say twelve inches wide, in the same manner that a door is made; turn it, cleated side down, and plane the upper face smooth. Finally, having laid on the leaf in position, so as to have the two cleats outside of the legs to prevent their slipping above, secure them below by a straight round piece laid from one pair of legs to the other, and notched on to them at their crossing or intersection.

This foot-bar might be still better made with its ends cut and fitted in as the pins or pivots of the cross-legs.

A table such as this can be taken apart, folded up and put away in a few minutes; and can be got ready for meals very expeditiously. It should be two feet four inches high.

THE CHAIRS.

Next to the table, in importance, come the chairs; and in the making of these, strength and comfort are the chief objects to be kept in view. The easiest mode of formation is that adopted for the table, just described; namely, two X's connected by a cross-rail at their intersecting points, and having a board-seat with four holes bored in it, into which the upper ends of the legs may be inserted; their tops being afterwards sawed off flush with the seat. This is now a stool. In order to make a chair of it (Fig. 42), a back must be constructed, and this is done in the following manner: Two sticks, about thirty inches long, are notched and secured to the legs (as shown in the figure), and are likewise notched, or let into the seat-board, and secured to a cross-piece, or round, at top. The dimensions of such chairs should be: From floor to top of seat-board, seventeen inches; the width sixteen inches; and height of back, from the seat-board, fifteen inches. Ash would be the best wood to use. The pieces should be an inch and a half or two inches thick, and should have the bark peeled off. The cross-legs should each be twenty-one inches long, and should cross one another at fourteen inches of their length from the bottom. The stays, or bracing-pieces, below the point of crossing, should be half or three-quarters of an inch thick, and be let into slots, on the inside of the legs, deep enough to insure their not working out.

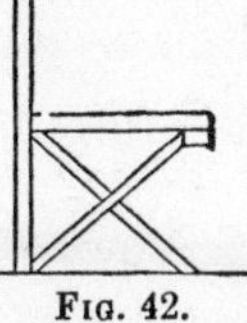

FIG. 42.

A SETTEE, OR SOFA.

As years go by, comforts gradually accumulate, and industry and energy begin to have their hard-earned reward. A new house is built, and handsome furniture provided for it. But even in the days of early struggle, there is no good reason why many little comforts, even in a rude way, should not be secured to the pioneer, and one of these comforts is the settee or sofa: which may readily be made by sawing a four-inch thick and two feet six long oak-stick into four parts of seven and a half inches long each. Two slots or grooves three inches long, each one inch deep and an inch wide, are to be cut with the chisel in each of these legs, from the top down. When the four legs are set up in position, two slats split or sawed from inch-boards eighteen inches long each, and two other slats five feet eight inches long each, are to be set tightly into the slots. Next a box is to be made of rough pine boards, and set upon these framed legs. This box is to be six feet long, one foot nine inches wide, and eight inches high. It is to be securely pinned or nailed down to the tops of the four legs. On front and rear of this box (next to the corners) half-inch thick boards, seven inches wide and eighteen inches long, are to be nailed on to the box, standing upright and rounded on the top. These form the supports for the arms of the sofa. Coarse canvas should now be tightly tacked on from one to the other of each of these arm-pieces, and hay should be carefully packed in them. Hay should likewise be packed in the box, until it is filled up. It should then be covered tightly with canvas, nailed on all around the outsides. A couple of bags of chaff, procured at the nearest flouring-mill, may now be carefully distributed over the canvas of both box and arms; and, lastly, the whole sofa should be covered with a neat pattern chintz.

THE BARREL-CHAIR.

Of all the makeshifts which necessity suggests, there is not one that so commends itself to the handy-man as the making an arm-chair out of a flour-barrel. It is indeed a most comfortable, as well as economic substitute for the cabinet-maker's design, and can, with a little taste in its upholstery, prove as neatly pleasant as the best. At sixteen or seventeen inches high, from the floor, saw the barrel half through, taking care to have a hoop immediately below the cut to secure the staves; and likewise a half round of hoop at the top of the back, carefully nailed on to the staves (Fig. 43). Below the level of the seat, and just above the hoop, bore, say twelve holes, through which pass a strong cord crossing and re-crossing. On this put a thick layer of hay, and over it lay a round cushion filled with chaff. The whole of this barrel-chair should be covered neatly with plaited chintz; the back being thinly lined with the same material filled in with cut hay. Of course, the barrel-chair can be used without the chintz, and we only mention it here as a desirable finish, which can be put on at any time.

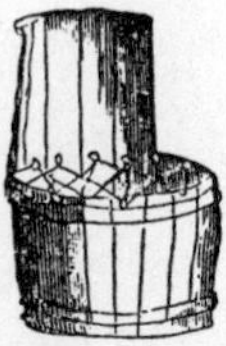

FIG. 43.

Foot-stools, and low seats for children, can be easily made out of candle-boxes, having no cover, filled with chaff, and covered with chintz.

Chaff will also make a good bed, as well as corn-husks, cat-tail, rushes, &c., until the good time coming shall bring a supply of goose-feathers or even *down*. But we are now advising for pioneer struggle, and not for the comforts which surely follow well-directed industry.

OUT-OF-DOOR SEATS.

No matter how small the dwelling, or how humble the beginning of the pioneer, there are a great many little ways in which he can add real luxuries to his homestead. Besides the indoor comforts, which a very small outlay of labor will secure to him, the handy-man can provide rustic seats out-of-doors for summer-evening enjoyment. These may be made of branches with the bark left on, and may be made pretty objects to look at, as well as to sit upon.

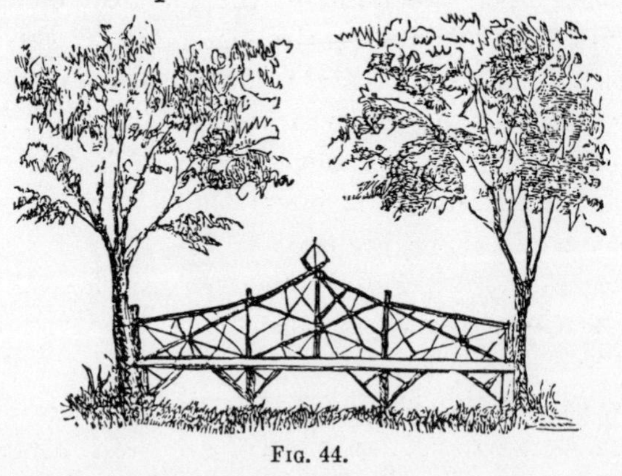

FIG. 44.

In making the clearing for the house, one goodly tree, at least, should be spared, say an old spreading oak, around the trunk of which might be constructed a rustic seat. These out-of-door seats should be permanent fixtures. For the legs and back, forked branches should be selected. The seat should be made of a slab, having its flat side turned up, and on this flat surface should be tacked small split branches in short lengths, laid with the split surface down. These slats should be uniform in size, and have their bark left on. They may be laid straight or diagonally, according to taste.

PAPER COVERING FOR WALLS AND ROOF.

WHERE coarse, strong, brown paper (made from straw) can be had, it is better as a material for covering roofs, and even walls (inside and out), than boards, shingles, or lath-and-plaster. This fact has so forced itself upon the mind of the practical man, that the manufacture of paper entirely suitable for this purpose has become very general; and it may now be had in any quantity, and at most reasonable prices, in all cities and villages throughout the land. But, in order to render the walls of a house sufficiently protective against cold air, it will be necessary [illegible] over the walls, inside and out, with surface-dressed boards, set close. The roof must likewise be so covered or sheathed.

In the case of the balloon-house, already described, it will be necessary to counter-sheathe the walls inside, and paper that only. But in frame construction, both sides of the walls must be sheathed with half-dressed boards, and the paper be applied on the inside and outside of the walls.

In order that my readers may have the benefit of a trustworthy gentleman's actual experience in this matter, I extract the following clear and very interesting remarks on the subject from a communication addressed to the Des Moines (Wisconsin) *Register*, by the Rev. W W. King:

"I want to give such information as I possess touching a matter of interest to the people, especially to those who desire to build comfortable houses, and whose pecuniary resources are limited. Anxious to secure a house for my family, and avoid paying one dollar a day rent, I began my preparations for building the 1st of last October, and the 1st of December moved into very comfortable winter quarters. After consulting with reliable parties and examining the material, I concluded to use the Rock River Company's building-paper, and it is of this experiment and its result that I desire more particularly to speak. I have used the paper on the outside instead of weather-boards or siding, on the roof instead of shingles, and on my inside walls throughout instead of plastering, and with results more than satisfactory. I have a home warmer than any plastered house I ever saw, and I have saved more than one-third of the expense of the old method. The erection of my building was in all respects an experiment; but I learned continually by trial, and could now improve on that experiment in many respects. I think I can give practical advice to those who desire to use this material.

"In the first place, as all the walls inside and outside are covered with common surface-dressed sheeting, the studding can be placed at least three feet apart, and the frame be stiffer than that of an ordinary building. Here there is a large saving of about one-half of the dimension-lumber required in a plastered building. I covered the outside walls and roof with sheeting. On the roof I used the plain paper, commencing at the eaves, and bending over the edge of the paper, and fastening to the edge of the roof-board with common-headed ten-ounce tacks, driven about one inch apart. The next course of paper was lapped over the joint four inches, taking pains to join the lap before laying with a heavy coat of mineral paint, then tacking as before, near the lower edge of the second course and through both. Care should be taken to tack the ends carefully. When the roof is thus laid, it should be covered with a heavy coat of mineral paint. Then have common inch-lumber cut into strips half an inch thick, and these strips laid up and down the roof from eight to ten inches apart, and fastened with shingle-nails; then cover all with mineral paint, and you have a roof that no wind or rain can penetrate, and which I am confident, if kept properly painted, will last an ordinary lifetime. On the outside walls, put the paper on perpendicularly, laying the edges of the courses together, and tacking once in four inches, and following with mineral paint. Then take common battens, paint them on the back side, lay once over the joints where the courses of paper meet, and once in the centre of each course; then paint as on the roof. Care should be taken outside and inside to lay the paper close to the window-frames and door-frames before the casings are set, so that the casings may cover all joints. On the inside I use common sheeting laid close, from the floor to the top of the back of a common chair; and from that point to the top and overhead I use strips of lumber from two to four inches wide — laid two feet apart — taking pains to have the edges of the paper meet the centre of a strip. The paper should be laid lengthwise on the strips, and the edges tacked each inch, with occasional tacks here and there through the centre of the paper. Then, when all is done, use for walls and ceiling any wall-paper to suit the taste. The walls and ceiling of the kitchen and pantry should be painted with mineral paint for a ground-work, and then any color to suit. Strips of paper or thin muslin pasted over the joints in the paper before painting, to hide the heads of tacks, will improve the appearance. For floors, use common surface-dressed lumber, and then cover with paper before carpeting. These plain directions will insure comfort and economy. The paper can be laid at a cost of from ten to twelve cents per yard. This includes everything, labor and material. Let us look at the net results of advantages:

"'*First.* You can build at any season of the year, finish your rooms to-day, and move into them to-morrow.

"'*Second.* You save a large expense in hauling lime and sand.

"'*Third.* You save largely in the amount and quality of labor required, as there is far less work, and it can mostly be done by common laborers.

"'*Fourth.* You save one-half of your dimension-lumber; and after the frame is up and the partitions set, it is less expense to prepare the walls for the paper than for plastering.

"'*Fifth.* You have a warmer house than can be built with lumber and mortar.'

"In conclusion, I believe the 'Paper-House' is a success, and will prove a great blessing and improvement in this climate, and especially to the thousands who must study economy."

In the matter of economy there is, no doubt, a great inducement to adopt this paper covering for walls; but at the same time I would urge the consideration of the stubborn fact that sheathing-boards will inevitably *shrink*, and that the paper attached to them will therefore split and fail of its purpose. However, there is yet a way of escape from this threatened mischief, and that is to lay the sheathing-boards diagonally (or crossways) instead of horizontally, or perpendicularly. The shrinking of the boards would then exercise little or no influence on the paper, as the joinings would be in different directions, and shrinkage always takes place at the sides, and not at the ends of the boards. And, even where the paper does split, it would be always an easy matter to re-paper on it; thus making the covering stronger, as well as renewing the appearance.

On the whole, I would recommend the use of this

paper covering; especially in our Southern or our Pacific localities. It is easily carried, cleanly in use, and causes no delay in the putting on. In many respects, it is far superior to lath-and-plaster; though in some it is inferior. For roofing it is very desirable; but, for the reasons already given, I would strongly recommend the boarding the roof diagonally.

This paper properly laid around the bases of chimney-shafts, would keep out the weather most effectually.

ARTIFICIAL HEARTHSTONE.

THE open fireplace to be found in every cottage requires a hearthstone, or its substitute, made of brick or tile. Supposing it a difficult matter to obtain these, then it becomes necessary to have recourse to the artificial stone, which can be very effectively made in the following manner: — Procure, at a butcher's, a quantity of bullocks' blood, and with one-tenth of its weight of quicklime reduced to powder by sprinkling water on it, mix up the mass, which may be thinned, if necessary, by water. Having paved the space to be occupied by the hearth, making the surface of the paving come within an inch of the surface of the floor; next, plaster over this paving with the blood-cement just mentioned, and make its surface perfectly smooth, and even with the surface of the floor of the room. Clean, sharp sand may be mixed in with the cement, and the paving should likewise have a coating of sand before the cement is laid on. This will prove a good, serviceable hearth, and can be polished to a high degree, if the woman of the house will take the pains to produce that pleasing effect, which the neat and ambitious housekeeper, no doubt, will cheerfully do. It must be borne in mind that it is the albumen, or gelatin, of the blood which is to be used. The fibrin, or clot, is to be thrown out.

Besides using this blood-composition for hearths, the whole inside of the fireplace might be plastered with it to very great advantage. The cellar (if there is to be one under the house) might be judiciously floored with it, as it becomes so very hard in time as to make a perfect stone floor all in one, being impervious to moisture and the inroads of vermin.

A cistern may be lined with this composition, and will be water-tight. And, in fact, its uses are very numerous, as it can be rubbed into rough canvas, and give an article fully equal to oil-cloth at a very trifling expense. It will be only necessary to stretch the canvas on a frame, and to rub the composition well into its surface until the texture becomes invisible. This makes a handy table-cover, being easily kept clean, and can be highly polished.

A cloth so prepared might be raised upon poles to shelter men from the intense heat of the sun when working away from other shelter, and would be found especially useful on the prairies.

As a water-proof cover for the wagon, or as a cloak for the man, it will alike make its utility felt.

Although bullock's blood is undoubtedly the best, yet any animal blood will be found to an extent useful for the purpose, according to the muscular development of the animal.

ADDITIONS AND IMPROVEMENTS.

IN course of time, the settler has so advanced as to be justifiable in making his family more comfortable in their homestead; and naturally turns his thoughts to an addition to its accommodation. This addition would be best situated next to the living-room, the back-door opening into it; and it should become the kitchen; such an arrangement would greatly conduce to the comfort and health of the family. The cooking, washing, &c., would thus be removed from the room in which the family sit and receive an occasional visitor; and in which, too, the man of the house and his wife sleep. Where fuel is in plenty, the additional fire for a kitchen would not be any extravagance, and even where fuel is not abundant, as on the prairies, it would be well to construct a brick hot-hearth with tile-top, having holes for pots and kettles (covered tightly when not in use). The flue from this could pass beneath the living-room floor into the chimney, which would, of course, have no fire-place. There should, however, be an independent flue to let off the smoke direct when the heat of summer would require the cutting off of the smoke-flue from the hot-hearth by a damper fixed for that purpose.

In connection with this proposed kitchen, it would be well to have a small pantry fitted up with shelves, nails, or pegs, and two bins to store the corn-meal and flour.

The well, or pump, should be conveniently located in the rear of this kitchen, so as to reduce the labor of water-carrying; and if the roof be of boards, shingles, or any material that will not cause rain-water to become impure in passing over it, a cistern may be sunk just outside the kitchen-wall, in which the rain that falls throughout the year can be collected and kept for use. Such a cistern must be made below ground to escape freezing in winter. In the deck, or top of it, a small pump might be inserted, the nose of which might be a tube extending into the kitchen, the handle being a rope attached to the sucker, and passing over a small wheel, or groove, and into the kitchen, having a stop larger than the hole in the kitchen-wall through which it passes.

All this can be easily executed by the handy-man, roughly to be sure, but nevertheless most acceptable to those whose labor it lessens.

Speaking of pumps, it would be very advisable that the cottager should make his own pump. There is nothing easier. Let him cut down a pine-tree suitable to his purpose, and having divested it of its superfluous branches and sawed it square off above the roots, with the help of neighbors place it *perfectly level*, and securely blocked, at say fifty inches high from the ground. Having procured a pump-borer's auger, to which he can attach the handle himself, if necessary, let him find the centre of the tree-trunk and chalk it. Next, let him set up in front of it a rest for his auger, which shall be a guide to direct the bore fairly.

If the pump is required to be so long that the shaft of the auger or borer will be too short to bore it through (working at both ends until the holes meet), then separate shafts must be bored, and be let into each other, by whittling the end of one outside and the end of the other inside. In such case, it would be a security to cover the joining with a stiff plaster of clay, and surround it with canvas or coarse cloth nailed tightly on. When the well-hole is very deep, this plan may have to be adopted more than once. For the cistern-pump, a small, straight bough, or a sapling, will be found sufficient.

The Immigrant may safely add this pump-auger to his other tools, for he will surely get enough work in the way of pump-boring from his neighbors to make it profitable as well as useful.

For animal drinking-troughs nothing can be better than the trunk of a large fallen tree, hollowed out with the axe. This trough should, of course, be convenient to the pump, for its supply of water, and there should be a hole and spigot near the bottom, to let off the water occasionally, and keep the supply clean and pure, for on this the health of the animals often depends.

WALKS IN YARD AND GARDEN.

THERE is no more necessary comfort within the reach of the humblest family, than hard, dry walks, by which to approach the front, or carry on the household business at the rear. Without these, the consequence of a rainy day or two, is dirty feet and dirty floor; not to speak of the discomfort of plashing through wet and mud, backward and forward, and especially in the evening, when daylight ceases to show you the places to be most avoided.

The making of walks, although apparently a very simple matter, requires much attention. Like the roof of a house, the walk should be formed with a view to the shedding of rain; and to do this, it becomes necessary to make the walk somewhat rounding, that is, highest at the centre, and inclining or sloping to the sides. This centre of the walk should be solidly constructed, so as to prevent any sinking there and formation of pools of rain-water; for if the centre once sinks in any part of the length of the walk, then will the whole walk be injured.

In order therefore to make a permanent walk, the site for it should be well beaten or trampled, and moderately large stones (the size of a man's fist) firmly bedded along the line of the centre, a foot and a half wide. On either side of this centre-belt, a course of smaller stones should be laid, inclining to the outer edge. Over the whole surface should then be spread a thick coat of sand or clay, and the entire walk should be well beaten, tramped, or, better still, rolled with a roller made out of a tree-trunk stripped of its bark, and cut square on the ends, having round pins firmly set in the centre at each end, and playing loose in a framed handle. If such a roller were axed on the surface, so as to exactly suit the required curve of the walk, it would be a great advantage, as it would save double rolling.

Lastly, the surface should be gravelled over, and the sides channelled, so as to give free passage to the rain-water which runs off the walk.

Where gravel or sand may not be plenty, the walks can be very well made of clay, taking due care to tramp and beat hard every layer, and especially the surface.

Where the ground is springy or spongy, trunks of trees must be laid lengthways, and sleepers pinned down across them securely at regular distances, the interspaces filled in with chips, stones, and clay. Whatever mode is had recourse to, it will be well worth the time and labor expended to make good and permanent walks.

A roadway to the barn would be a convenience which would force its necessity on the thriving man; and it would be well, if from the very outset he should lay it out, and form it even by slow degrees.

USEFUL INFORMATION.

THE following tables will be found correct, and very often useful for reference:

A box 24 inches square and 16 inches deep, will contain a barrel very nearly.

A box 7 by 8 inches and 4½ inches deep, holds a gallon; 40 such boxes full would fill a barrel.

In measuring or making a circular cistern or well, 2 feet in diameter and 10 inches deep, will contain 19 gallons.

— 2½ feet in diameter and 10 inches deep, will hold 30 gallons.

— 3 feet in diameter and 10 inches in depth = 44 gallons.

If the well or cistern be 10 feet in depth, it will have a capacity of exactly twelve times these quantities.

Corn may be measured thus: Two cubic feet of good, sound, dry corn in the ear will make one bushel of shelled corn, so that by measuring the length, breadth, and height of a corn-crib inside, and multiplying them together, then dividing the amount by 2, the number of bushels of shelled corn in the crib will be ascertained.

Fire-wood, for the market, is sold by the cord of 128 solid feet; that is, 4 feet long, 4 feet broad, and 8 feet high.

By observing the following measure, the amount of land occupied by fences, roads, &c., may be easily calculated. It is this: A strip of one mile long (1,760 yards) and 8 feet 3 inches wide will cover two acres. If half a mile long and 8 feet 3 inches wide = 1 acre.

A square acre measures 209 feet on each side. It covers 4,840 square yards, and is the same as an English acre.

640 acres make a section of Government land.

160 acres, or half a mile square, is a quarter section, which is the amount granted as soldiers' bounty land.

36 sections, or 144 soldiers' bounty grants, will measure 36 miles square, and is called a township.

The townships are laid off in solid, or square, blocks. Section No. 1 begins at the north-east corner of each block, or township; and the others follow, north-westwardly, to No. 6, and on the next line they count back to the north-east, and so on, backward and forward, until No. 36 is reached in the north-east corner.

The section No. 16 is appropriated by the Government as the Township School Section, being nearly central.

Glue.—As glue is a thing which is often used about a house, and often put aside after use, it always happens that the remainder in the glue-pot is so hard when required again that it takes much time and trouble to remelt it. A few simple instructions, therefore, upon the subject may not be amiss to the handy-man:—It must be understood that to melt glue properly, two vessels must be used, one within the other; the difference in size being such as to admit of the inner one being surrounded with water contained in the outer one. The glue is placed in the inner vessel, and when the water is put in the outer one, it is set over the fire to boil, and the inner one, containing the glue, is placed in it. When the glue has been used, and the remainder is about to be set aside, it will be found useful to pour in on the glue a little cold water. When re-heating this glue at any time, this water is permitted to remain, and the same process gone through as before. The glue will easily dissolve as the water heats. If this precaution of pouring water on the glue when setting it aside is not taken, it will inevitably become hard, and give much trouble in re-melting. Some practical men (cabinet-makers especially) steep their glue for eight hours in as much cold water as will just cover it, previously breaking it up into small pieces.

A Turning-Lathe.—The handy-man could not possibly find for an occasional leisure hour a more profitable occupation than that afforded by a turning-lathe. All the primitive makeshifts, which have been described and suggested in the pages of this little volume, could then be vastly improved upon, and many conveniences and comforts added.

The cost of such a useful article to him would be trifling, compared with the pleasure which the handy-man would derive from it. In fact, with the proper use of his brains, he could make up one himself which would answer until he could afford to supersede it by one of a more mechanic form and action.

Take three pieces of two-inch plank, each ten inches wide and thirty inches long, and cleat them together crosswise. Then find the centre, and stick a brad-awl, or a nail, in it; from this stretch a string thirty inches long, at the end of which tie a pencil, or chalk, and mark out a circle. Now saw around on this line-mark, and the driving-wheel is roughed out. At the centre chisel out a square hole, which may first be bored with the auger, or a red-hot poker. At the blacksmith's get a bar of inch-round iron, say three feet long, and get him to beat it out, and form it so as to have a curve, or V, in the middle, about eight inches deep. Insert one end of this iron bar in the centre of the wheel passing through it, say three or four inches, and resting in a hole made for it in an upright piece; the other end to rest in a similar hole in an upright piece at the other extremity, and both exactly of the same height. Directly under the V fix a treadle working on pivots in two stakes firmly driven in the ground; connect the end of this treadle with the V at its sharp angle with strong wire of a length to permit the V to make a perfect revolution. This turns the wheel with the aid of a long ash-springer attached at one end to the ceiling, and at the other by a cord, to the spindle around which it is turned three or four times. A leather belt or thong is made to pass over the driving-wheel and around a small wheel directly above it, through the centre of which is the spindle supported at four points; that is, at its ends and at the middle, where it is jointed, and where bits can be inserted to hold the piece to be turned. Lastly, a *rest* is constructed directly in front of this joint. On this the turner rests his hands while holding the cutter to the revolving block, his foot at the same time working the treadle assisted by the action of the spring above his head.

Such tools as may be most needed for Turning can be had at any hardware store.

Before putting on the running band, it would be well to turn a ploughed groove in the wheel to receive it and prevent any occasional slipping out.

CONCLUDING REMARKS.

IT must be distinctly understood, that, in the preceding pages, the object sought to be kept in view, was advising a working-man who had immigrated to any location distant from civilization, or society, to found a homestead for his family, and adapt his means to his purpose most advantageously. Taste he may possess, but the application of it must always be made subservient to utility and economy.

Where several families are migrating in company, it would be very desirable a carpenter or two should be among them. For, although masons or bricklayers might be very useful in a small pioneer colony, such as that we have under consideration, they may be easier dispensed with than can the carpenter. But in the foregoing pages we have supposed the total absence of all mechanic skill, and given the plainest, if not the best, directions to the man of the family, by which, if he be in the least handy, he can turn to account the things within his reach, and make comfort grow out of inconvenience.

The pioneer settler cannot be too careful to provide himself with the tools, of which we have given a list, or at least with those which his own sense will tell him he will most require. A couple of kegs of assorted nails will never be amiss. A large and a small paint-brush, with a gallon of boiled linseed-oil, and a quart of turpentine, will enable him to secure his door- and window-frames from the consequences of their exposure to the weather, as well as to make the interior neat and comfortable to the eye.

Prudence and forethought will prompt him to buy at the village nearest to his destination, all things that he will be likely to want most, as it is often very difficult to procure them afterwards in a new and sparsely peopled section, where roads, if there are any, are of the roughest and most impassable sort.

There is great economy in taking care of tools; and for this purpose it will be necessary to keep them in a closet or box; and, in order to prevent rusting, it will be advisable to smear them with grease on the steel or iron parts. It would be well also to keep sharp edges on tools, so as to have them ready for service on an emergency.

With these few remarks, we take our leave of the immigrant settler, wishing him a full supply of that patience and perseverance which are so sure to make industry and energy completely triumphant.

INDEX.